T0250596

Lecture Notes in Computer Science 1194

Edited by G. Goos, J. Hartmanis and J. van Leeuwen

Advisory Board: W. Brauer D. Gries J. Stoer

Springer

Berlin
Heidelberg
New York
Barcelona
Budapest
Hong Kong
London
Milan
Paris
Santa Clara
Singapore
Tokyo

Moshe Sipper

Evolution of Parallel Cellular Machines

The Cellular Programming Approach

 Springer

Series Editors

Gerhard Goos, Karlsruhe University, Germany

Juris Hartmanis, Cornell University, NY, USA

Jan van Leeuwen, Utrecht University, The Netherlands

Author

Moshe Sipper
Swiss Federal Institute of Technology
Logic Systems Laboratory, IN-Ecublens
CH-1015 Lausanne, Switzerland
E-mail: Moshe.Sipper@di.epfl.ch

Cataloging-in-Publication data applied for

Die Deutsche Bibliothek - CIP-Einheitsaufnahme

Sipper, Moshe:
Evolution of parallel cellular machines : the cellular
programming approach / Moshe Sipper. - Berlin ; Heidelberg ;
New York ; Barcelona ; Budapest ; Hong Kong ; London ;
Milan ; Paris ; Santa Clara ; Singapore ; Tokyo : Springer, 1997
  (Lecture notes in computer science ; 1194)
   ISBN 3-540-62613-1
NE: GT

CR Subject Classification (1991): F.1, C.1.2, F.2.2, D.1.3, I.2.6, J.3

ISSN 0302-9743
ISBN 3-540-62613-1 Springer-Verlag Berlin Heidelberg New York

This work is subject to copyright. All rights are reserved, whether the whole or part of the material is
concerned, specifically the rights of translation, reprinting, re-use of illustrations, recitation, broadcasting,
reproduction on microfilms or in any other way , and storage in data banks. Duplication of this publication
or parts thereof is permitted only under the provisions of the German Copyright Law of September 9, 1965,
in its current version, and permission for use must always be obtained from Springer -Verlag. Violations are
liable for prosecution under the German Copyright Law .

© Springer-Verlag Berlin Heidelberg 1997
Printed in Germany

Typesetting: Camera-ready by author
SPIN 10549006   06/3142 – 5 4 3 2 1 0   Printed on acid-free paper

To see a world in a grain of sand
And a heaven in a wild flower,
Hold infinity in the palm of your hand
And eternity in an hour.

William Blake, *Auguries of Innocence*

Preface

There is grandeur in this view of life, with its several powers, having been originally breathed into a few forms or into one; and that, whilst this planet has gone cycling on according to the fixed law of gravity, from so simple a beginning endless forms most beautiful and most wonderful have been, and are being, evolved.

Charles Darwin, *The Origin of Species*

Natural evolution has "created" a multitude of systems in which the actions of simple, locally-interacting components give rise to coordinated global information processing. Insect colonies, cellular assemblies, the retina, and the immune system, have all been cited as examples of systems in which *emergent computation* occurs. This term refers to the appearance of global information-processing capabilities that are not explicitly represented in the system's elementary components or in their interconnections.

The *parallel cellular machines* "designed" by nature exhibit striking problem-solving capacities, while functioning within a dynamic environment. The central question posed in this volume is whether we can mimic nature's achievement, creating artificial machines that exhibit characteristics such as those manifest by their natural counterparts. Clearly, this ambitious goal is yet far off, however, our intent is to take a small step toward it.

The first issue that must be addressed concerns the basic design of our system, namely, we must choose a viable *machine model*. We shall present a number of systems in this work, which are essentially generalizations of the well-known cellular automata (CA) model. CAs are dynamical systems in which space and time are discrete. A cellular automaton consists of an array of cells, each of which can be in one of a finite number of possible states, updated synchronously in discrete time steps, according to a local, *identical* interaction rule. CAs exhibit three notable features, namely, massive parallelism, locality of cellular interactions, and simplicity of basic components (cells). Thus, they present an excellent point of departure for our forays into parallel cellular machines.

Having chosen the machine model, we immediately encounter a major problem common to such local, parallel systems, namely, the painstaking task one is faced with in designing them to exhibit a specific behavior or solve a particular problem. This results from the local dynamics of the system, which renders the design of local interaction rules to perform global computational tasks extremely arduous. Aiming to learn *how to design* such parallel cellular machines, we turn to nature, seeking inspiration in the process of evolution. The idea of applying the biological principle of natural evolution to artificial systems, introduced more than three decades ago, has seen impressive growth in the past decade. Usually grouped under the term *evolutionary algorithms* or *evolutionary computation*, we find the domains of genetic algorithms, evolution strategies, evolutionary programming, and genetic programming. In this volume we employ artificial evolution, based on the genetic-algorithms approach, to evolve ("design") parallel cellular machines.

The book consists of three parts. After presenting the overall framework in Chapter 1, including introductory expositions of cellular automata and genetic algorithms, we move on to Chapter 2, the first part of this volume. We focus on *non-uniform cellular automata*, the machine model which serves as a basis for the succeeding parts. Such automata function in the same way as uniform ones, the only difference being in the local cellular interaction rules that need not be identical for all cells. In Chapter 2 we investigate the issue of universal computation, namely, the construction of machines, embedded in cellular space, whose computational power is equivalent to that of a universal Turing machine. This is carried out in the context of 2-dimensional, 2-state, 5-neighbor cellular space, that is not computation universal in the uniform case. We show that non-uniform CAs can attain universal computation using systems that are simpler than previous ones and are *quasi*-uniform, meaning that the number of distinct rules is extremely small with respect to rule-space size, distributed such that a subset of dominant rules occupies most of the grid. The final system presented is minimal, with but two distinct rules. Thus, we demonstrate that simple, non-uniform CAs comprise viable parallel cellular machines.

Chapter 3, the second part of this volume, investigates issues pertaining to artificial life (ALife). We present a modified non-uniform CA model, with which questions of evolution, adaptation, and multicellularity are addressed. Our ALife system consists of a 2-dimensional grid of interacting "organisms" that may evolve over time. We first present designed multicellular organisms that display several interesting behaviors, including reproduction, growth, and mobility. We then turn our attention to evolution in various *environments*, including an environment where competition for space occurs, an IPD (Iterated Prisoner's Dilemma) environment, a spatial-niches environment, and a temporal-niches one. Several measures of interest are introduced, enabling us to glean the evolutionary process' inner workings. This latter point is a prime advantage of ALife models, namely, our enhanced investigative power in comparison to natural systems.

Our main conclusion from this part is that non-uniform CAs and their extensions comprise austere yet versatile models for studying ALife phenomena. It is hoped that the development of such ALife models will serve the two-fold goal of: (1) increasing our understanding of biological phenomena, and (2) enhancing our insight into artificial systems, thereby enabling us to improve their performance. ALife research opens new doors, providing us with novel opportunities to explore issues such as adaptation, evolution, and emergence, that are central both in natural environments as well as man-made ones.

In the third and main part of this volume, consisting of Chapters 4 through 8, we focus on the evolution of parallel cellular machines that solve complex, global computational tasks. In Chapter 4 we introduce the basic approach, denoted *cellular programming*, whereby a non-uniform CA *locally coevolves* to solve a given task. Our approach differs from the standard genetic algorithm, where a population of *independent* problem solutions *globally* evolves. We demonstrate the viability of our methodology by conducting an in-depth study of two global computational problems, density and synchronization, which are successfully solved by coevolved machines. In Chapter 5 we describe a number of additional computational tasks, motivated by real-world problems, for which parallel cellular machines were evolved via cellular programming. These tasks include counting, ordering, boundary computation, thinning, and random number generation, suggesting possible application domains for our systems.

Though most of the results described in this volume have been obtained through software simulation, a prime motivation of our work is the attainment of "evolving ware," *evolware*, with current implementations centering on hardware, while raising the possibility of using other forms in the future, such as *bioware*. This idea, whose origins can be traced to the cybernetics movement of the 1940s and the 1950s, has recently resurged in the form of the nascent field of bio-inspired systems and evolvable hardware. The field draws on ideas from evolutionary computation as well as on recent hardware developments. Chapter 6 presents the "firefly" machine, an evolving, online, autonomous hardware system that implements the cellular programming algorithm, thus demonstrating that *evolware* can indeed be attained.

Most classical software and hardware systems, especially parallel ones, exhibit a very low level of fault tolerance, i.e., they are not resilient in the face of errors; indeed, where software is concerned, even a single error can often bring an entire program to a grinding halt. Future computing systems may contain thousands or even millions of computing elements. For such large numbers of components, the issue of resilience can no longer be ignored since faults will be likely to occur with high probability. Chapter 7 looks into the issue of fault tolerance, examining the resilience of our evolved systems when operating under faulty conditions. We find that they exhibit graceful degradation in performance, able to tolerate a certain level of faults.

A fundamental property of the original CA model is its standard, homogeneous connectivity, meaning that the cellular array is a regular grid, all cells connected in exactly the same manner to their neighbors. In Chapter 8 we study *non-standard* connectivity architectures, showing that these entail definite performance gains, and that, moreover, one can evolve the architecture through a two-level evolutionary process, in which the local cellular interaction rules evolve concomitantly with the cellular connections.

Our main conclusion from the third part is that parallel cellular machines can attain high performance on complex computational tasks, and that, moreover, such systems can be evolved rather than designed. Chapter 9 concludes the volume, presenting several possible avenues of future research.

Parallel cellular machines hold potential both scientifically, as vehicles for studying phenomena of interest in areas such as complex adaptive systems and artificial life, as well as practically, showing a range of potential future applications, ensuing the construction of systems endowed with evolutionary, reproductive, regenerative, and learning capabilities. We hope this volume sheds light on the behavior of such machines, the complex computation they exhibit, and the application of artificial evolution to attain such systems.

Acknowledgments

I thank you for your voices: thank you:
Your most sweet voices.
William Shakespeare, *Coriolanus*

It is a pleasure to acknowledge the assistance of several people with whom I collaborated. Daniel Mange, Eduardo Sanchez, and Marco Tomassini, from the Logic Systems Laboratory at the Swiss Federal Institute of Technology in Lausanne, were (and still are!) a major source of inspiration and energy. Our animated discussions, the stimulating brainstorming sessions we held, and their penetrating insights, have seeded many a fruit, disseminated throughout this volume. I thank Eytan Ruppin from Tel Aviv University, with whom it has always been a joy to work, for having influenced me in more ways than one, and for his steadfast encouragement during the waning hours of my research. Pierre Marchal from the Centre Suisse d'Electronique et de Microtechnique was a constant crucible of ideas, conveyed in his homely, jovial manner, and I have always been delighted at the opportunity to collaborate with him. The Logic Systems Laboratory has provided an ideal environment for research, combining both keen minds and lively spirits. I thank each and every one of its members, and am especially grateful to Maxime Goeke, Andrés Pérez-Uribe, André Stauffer, Mathieu Capcarrere, and Olivier Beuret. I am grateful to Melanie Mitchell from the Santa Fe Institute for her many valuable comments and suggestions. I thank Hezy Yeshurun from Tel Aviv University for his indispensable help at a critical point in my research. I am grateful to Hans-Paul Schwefel from the University of Dortmund for reviewing this manuscript, offering helpful remarks and suggestions for improvements. I thank Alfred Hofmann and his team at Springer-Verlag, without whom this brainchild of mine would have remained just that – a pipe dream. Finally, last but not least, I am grateful to my parents, Shoshana and Daniel, for bequeathing and believing.

Lausanne, December 1996   Moshe Sipper

Contents

Chapter 1

Introduction

> The White Rabbit put on his spectacles. "Where shall I begin, please your Majesty?" he asked.
> "Begin at the beginning," the King said, very gravely, "and go on till you come to the end: then stop."
>
> Lewis Carroll, *Alice's Adventures in Wonderland*

1.1 Prologue

Natural evolution has "created" a multitude of systems in which the actions of simple, locally-interacting components give rise to coordinated global information processing. Insect colonies, cellular assemblies, the retina, and the immune system, have all been cited as examples of systems in which *emergent computation* occurs (Langton, 1989; Langton et al., 1992). This term refers to the appearance of global information-processing capabilities that are not explicitly represented in the system's elementary components or in their interconnections (Das et al., 1994). As put forward by Steels (1994), a system's behavior is emergent if it can only be defined using descriptive categories that are not necessary to describe the behavior of the constituent components. As a simple example, consider the regularities observed in the collective behavior of many molecules, entailing new categories like temperature and pressure.

The *parallel cellular machines* "designed" by nature exhibit striking problem-solving capabilities, while functioning within a dynamic environment. The central question posed in this volume is whether we can mimic nature's achievement, creating artificial machines that exhibit characteristics such as those manifest by their natural counterparts. Clearly, this ambitious goal is yet far off, however, our intent is to take a small step toward it.

Our interest lies with systems in which many locally connected processors, with no central control, interact in parallel to produce globally-coordinated behavior that is more powerful than can be done by individual components or linear combinations of components. The first issue that must be addressed is that of the basic design of our system, namely, we must choose a viable *machine model*. We

shall present a number of systems in this volume, which are essentially generalizations of the well-known cellular automata (CA) model. CAs, described in detail
in the next section, exhibit three notable features, namely, massive parallelism,
locality of cellular interactions, and simplicity of basic components (cells). They
perform computations in a distributed fashion on a spatially-extended grid; as
such, they differ from the standard approach to parallel computation, in which
a problem is split into independent sub-problems, each solved by a different processor, later to be combined in order to yield the final solution. CAs suggest a
new approach, in which complex behavior arises in a bottom-up manner from
non-linear, spatially-extended, local interactions (Mitchell et al., 1994b). Thus,
the CA model presents an excellent point of departure for our forays into parallel
cellular machines.

Upon settling the first issue, choosing the machine model, we immediately
encounter a major problem common to such local, parallel systems, namely, the
painstaking task one is faced with in designing them to exhibit a specific behavior
or solve a particular problem. This results from the local dynamics of the system,
which renders the design of local interaction rules to perform global computational tasks extremely arduous. Automating the design (programming) process
would greatly enhance the viability of such systems (Mitchell et al., 1994b). Thus,
the second issue is that of *how to design* such parallel cellular machines. Again,
we turn to nature, seeking inspiration in the process of evolution.

The idea of applying the biological principle of natural evolution to artificial
systems, introduced more than three decades ago, has seen impressive growth
in the past decade. Usually grouped under the term *evolutionary algorithms*
or *evolutionary computation*, we find the domains of genetic algorithms, evolution strategies, evolutionary programming, and genetic programming. Central to
all these different methodologies is the idea of solving problems by evolving an
initially random population of possible solutions, through the application of "genetic" operators, such that in time "fitter" (i.e., better) solutions emerge (Bäck,
1996; Michalewicz, 1996; Mitchell, 1996; Schwefel, 1995; Fogel, 1995; Koza, 1992;
Goldberg, 1989; Holland, 1975). We shall employ artificial evolution to evolve
("design") parallel cellular machines.

The issues we investigate pertain to the fields of complex adaptive systems and
artificial life (ALife). The former is concerned with understanding the laws and
mechanisms by which global behavior emerges in locally-interconnected systems
of simple parts. As noted above, these abound in nature at various levels of
organization, including the physical, chemical, biological, and social levels. The
field has experienced rapid growth in the past few years (Coveney and Highfield,
1995; Kaneko et al., 1994; Kauffman, 1993; Pagels, 1989).

Artificial life is a field of study devoted to understanding life by attempting
to abstract the fundamental dynamical principles underlying biological phenomena, and recreating these dynamics in other physical media, such as computers,
making them accessible to new kinds of experimental manipulation and testing

(Langton, 1992b). Artificial life represents an attempt to vastly increase the role
of synthesis in the study of biological phenomena (Langton, 1994; see also Sipper,
1995a; Levy, 1992). As noted by Mayr (1982): "The question of what the major
current problems of Biology are cannot be answered, for I do not know of a single biological discipline that does not have major unresolved problems ... Still,
the most burning and as yet most intractable problems are those that involve
complex systems."

We commence our study of parallel cellular machines with an exposition of
cellular automata and genetic algorithms. These serve as a basis for our work,
and are presented in the following two sections.

1.2   Cellular automata

1.2.1   An informal introduction

Cellular automata (CA) were originally conceived by Ulam and von Neumann
in the 1940s to provide a formal framework for investigating the behavior of
complex, extended systems (von Neumann, 1966). CAs are dynamical systems
in which space and time are discrete. A cellular automaton consists of an array
of cells, each of which can be in one of a finite number of possible states, updated
synchronously in discrete time steps, according to a local, identical interaction
rule. The state of a cell at the next time step is determined by the current states
of a surrounding neighborhood of cells (Wolfram, 1984b; Toffoli and Margolus,
1987).

The cellular array (grid) is n-dimensional, where $n = 1, 2, 3$ is used in practice;
in this volume we shall concentrate on $n = 1, 2$, i.e., one- and two-dimensional
grids. The *identical* rule contained in each cell is essentially a finite state machine,
usually specified in the form of a *rule table* (also known as the *transition function*),
with an entry for every possible neighborhood configuration of states. The *cellular
neighborhood* of a cell consists of the surrounding (adjacent) cells. For one-
dimensional CAs, a cell is connected to r local neighbors (cells) on either side, as
well as to itself, where r is a parameter referred to as the radius (thus, each cell has
$2r + 1$ neighbors). For two-dimensional CAs, two types of cellular neighborhoods
are usually considered: 5 cells, consisting of the cell along with its four immediate
nondiagonal neighbors, and 9 cells, consisting of the cell along with its eight
surrounding neighbors. When considering a finite-sized grid, spatially periodic
boundary conditions are frequently applied, resulting in a circular grid for the
one-dimensional case, and a toroidal one for the two-dimensional case.

As an example, let us consider the parity rule (also known as the XOR rule)
for a 2-state, 5-neighbor, two-dimensional CA. Each cell is assigned a state of 1
at the next time step if the parity of its current state and the states of its four
neighbors is odd, and is assigned a state of 0 if the parity is even (alternatively,
this may be considered a modulo-2 addition). The rule table consists of entries

CNESW	S_{next}	CNESW	S_{next}	CNESW	S_{next}	CNESW	S_{next}
00000	0	01000	1	10000	1	11000	0
00001	1	01001	0	10001	0	11001	1
00010	1	01010	0	10010	0	11010	1
00011	0	01011	1	10011	1	11011	0
00100	1	01100	0	10100	0	11100	1
00101	0	01101	1	10101	1	11101	0
00110	0	01110	1	10110	1	11110	0
00111	1	01111	0	10111	0	11111	1

Table 1.1. Parity rule table. CNESW denotes the current states of the center, north, east, south, and west cells, respectively. S_{next} is the cell's state at the next time step.

of the form:

$$\boxed{1}\,\boxed{1}\,\boxed{0} \mapsto 1$$

This means that if the current state of the cell is 1 and the states of the north, east, south, and west cells are $0, 0, 1, 1$, respectively, then the state of the cell at the next time step will be 1 (odd parity). The rule is completely specified by the rule table given in Table 1.1. Figure 1.1 demonstrates patterns that are produced by the parity CA.

The CA model is both *general* and *simple* (Sipper, 1995c). Generality implies two things: (1) the model supports universal computation (Chapter 2), and (2) the basic units encode a general form of local interaction rather than some specialized action (Chapter 3). Simplicity implies that the basic units of interaction are "modest" in comparison to Turing machines. If we imagine a scale of complexity, with one end representing Turing machines, then the other end represents simple machines, e.g., finite state automatons. The CA model is one of the simplest, general models available. From the point of view of parallel cellular machines, CAs exhibit three notable features, namely, massive parallelism, locality of cellular interactions, and simplicity of basic components (cells).

1.2.2 Formal definitions

The following section provides formal definitions concerning the CA model, due to Codd (1968). In the interest of simplicity we concentrate on two-dimensional CAs (the general n-dimensional case can be straightforwardly obtained). Let I denote the set of integers. To obtain a cellular space we associate with the set $I \times I$:

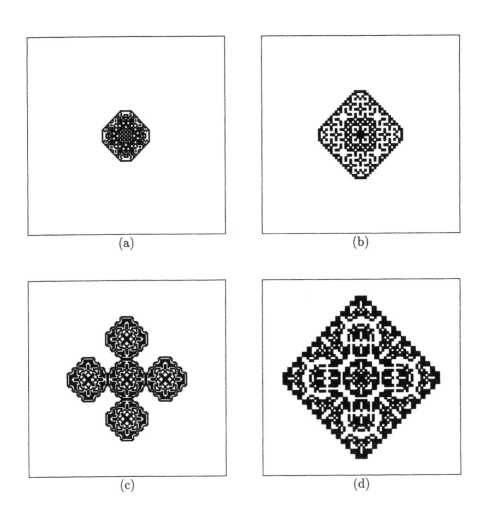

Figure 1.1. Patterns produced by the parity rule, starting from a 20×20 rectangular pattern. White squares represent cells in state 0, black squares represent cells in state 1. (a) after 30 time steps ($t = 30$), (b) $t = 60$, (c) $t = 90$, (d) $t = 120$.

1. The *neighborhood function* g: $I \times I \to 2^{I \times I}$, defined by

$$g(\alpha) = \{\alpha + \delta_1, \alpha + \delta_2, \ldots, \alpha + \delta_n\},$$

for all $\alpha \in I \times I$, where $\delta_i \in I \times I$ $(i = 1, 2, \ldots, n)$, is fixed.

2. The finite automaton (V, v_0, f), where V is the set of *cellular states*, v_0 is a distinguished element of V called the *quiescent state*, and f is the local transition function from n-tuples of elements of V into V. The function f is subject to the restriction

$$f(v_0, v_0, \ldots, v_0) = v_0.$$

Essentially, there is a (two-dimensional) grid of interconnected cells, each containing an *identical copy* of the finite automaton (V, v_0, f).[1] The state $v^t(\alpha)$ of a cell α at time t is precisely the state of its associated automaton at time t. Each cell α is connected to the n neighboring cells $\alpha + \delta_1, \alpha + \delta_2, \ldots, \alpha + \delta_n$. In all that follows we assume that one of the neighbors of α is α itself and we adopt the convention that $\delta_1 = (0, 0)$.

The *neighborhood state function* h^t: $I \times I \to V^n$ is defined by

$$h^t(\alpha) = (v^t(\alpha), v^t(\alpha + \delta_2), \ldots, v^t(\alpha + \delta_n)).$$

Now we can relate the neighborhood state of a cell α at time t to the cellular state of that cell at time $t + 1$ by

$$f(h^t(\alpha)) = v^{t+1}(\alpha).$$

The function f is referred to as the CA *rule* and is usually given in the form of a *rule table*, specifying all possible pairs of the form $(h^t(\alpha), v^{t+1}(\alpha))$. Such a pair is termed a *transition* or *rule-table entry*. When convenient, we omit the time superscript t from h^t.

An allowable assignment of states to all cells in the space is called a *configuration*. Thus, a configuration is a function c from $I \times I$ into V, such that

$$\{\alpha \in I \times I \mid c(\alpha) \neq v_0\}$$

is finite. Such a function is said to have *finite support relative* to v_0, and the set above is denoted $sup(c)$. This is in accordance with von Neumann who restricted attention to the case in which all cells except a finite number are initially in the quiescent state (von Neumann, 1966). Since $f(v_0, v_0, \ldots, v_0) = v_0$ (i.e., a cell whose neighborhood is entirely quiescent remains quiescent), at every time step all cells except a finite number are in the quiescent state.

---

[1] For non-uniform cellular spaces, different cells may contain different transition functions, i.e., f depends on α, f_α (see next section).

We now define the *global transition function* F. Let C be the class of all configurations for a given cellular space. Then F is a function from C into C defined by

$$F(c)(\alpha) = f(h(\alpha))$$

for all $\alpha \in I \times I$. Given any initial configuration c_0, the function F determines a sequence of configurations

$$c_0, c_1, \ldots, c_t, \ldots$$

where

$$c_{t+1} = F(c_t)$$

for all t.

The configurations c, c' are disjoint if $sup(c) \cap sup(c') = \phi$. If c, c' are disjoint configurations, their *union* is defined by

$$(c \cup c')(\alpha) = \begin{cases} c(\alpha) & \text{if } \alpha \in sup(c) \\ c'(\alpha) & \text{if } \alpha \in sup(c') \\ v_0 & \text{otherwise} \end{cases}$$

A natural metric to associate with any cellular space based on $I \times I$ is the so-called city-block metric τ, defined by

$$\tau(\alpha, \beta) = \mid x_\alpha - x_\beta \mid + \mid y_\alpha - y_\beta \mid,$$

where $\alpha = (x_\alpha, y_\alpha)$ and $\beta = (x_\beta, y_\beta)$.

The metric τ is extended to finite sets of cells by

$$\tau(P, Q) = \max_{\alpha \in P, \beta \in Q} \tau(\alpha, \beta),$$

where P, Q are any finite sets of cells. We call $\tau(P, P)$ the *diameter* of P and abbreviate it $dia(P)$. Note that the extended function is not a true metric since in general $\tau(P, P) \neq 0$.

The extended function can now be applied to configurations as follows:

$$\tau(c, d) = \tau(sup(c), sup(d))$$

and

$$dia(c) = dia(sup(c)),$$

where c, d are any configurations.

We can now go on to define various notions related to propagation. Given a cellular space $(I \times I, n, V, v_0, f)$ we have observed above that an initial configuration c_0 determines a sequence of configurations. We shall call such a sequence a *propagation* and denote it by $<c_0>$.

A propagation $<c>$ is *bounded* if there exists an integer K such that for all t

$$\tau(c, F^t(c)) < K.$$

Otherwise, $< c >$ is an *unbounded propagation*. $F^t(c)$ denotes the result of t applications of the global transition function F to configuration c.

Suppose propagation $< c >$ is unbounded. It may be possible to limit its growth by adjoining to configuration c at time 0 some other (disjoint) configuration. Accordingly, we define a *boundable propagation* $<c>$ as one for which there exists a disjoint configuration d such that $c \cup d$ is bounded. In this case we say that d *bounds* c. If no d bounds a given c, $<c>$ is said to be an *unboundable propagation*.

1.2.3 Non-uniform CAs

The basic model we employ in this volume is an extension of the original CA model, termed *non-uniform cellular automata*. Such automata function in the same way as uniform ones, the only difference being in the cellular rules that need not be identical for all cells. Note that non-uniform CAs share the basic "attractive" properties of uniform ones (massive parallelism, locality of cellular interactions, simplicity of cells). From a hardware point of view we observe that the resources required by non-uniform CAs are identical to those of uniform ones since a cell in both cases contains a rule. Although simulations of uniform CAs on serial computers may optimize memory requirements by retaining a single copy of the rule, rather than have each cell hold one, this in no way detracts from our argument. Indeed, one of the primary motivations for studying CAs stems from the observation that they are naturally suited for hardware implementation with the potential of exhibiting extremely fast and reliable computation that is robust to noisy input data and component failure (Gacs, 1985). As noted in the preface, one of our goals is the attainment of "evolving ware," *evolware*, with current implementations centering on hardware, while raising the possibility of using other forms in the future, such as *bioware* (see Chapter 6).

Note that the original, uniform CA model is essentially "programmed" at an extremely low level (Rasmussen et al., 1992); a *single* rule is sought that must be universally applied to all cells in the grid, a task that may be arduous even if one takes an evolutionary approach. For non-uniform CAs search-space sizes (Section 1.3) are vastly larger than with uniform CAs, a fact that initially seems as an impediment. However, as we shall see in this volume, the model presents novel dynamics, offering new and interesting paths in the study of complex adaptive systems and artificial life.

1.2.4 Historical notes

This section presents some historical notes and references concerning CAs and is by no means a complete account of CA history. We shall present several other results, that are more closely linked with our own work, in the relevant chapters ahead (for more detailed accounts, refer to Gutowitz, 1990; Toffoli and Margolus, 1987; Preston, Jr. and Duff, 1984).

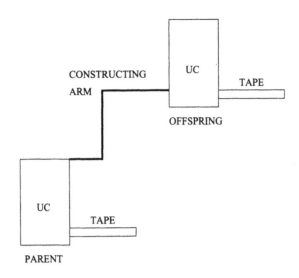

Figure 1.2. A schematic diagram of von Neumann's self-reproducing CA, essentially a universal constructor (UC) that is given, as input, its own description.

The CA model was originally introduced in the late 1940s by Ulam and von Neumann and used extensively by the latter to study issues related with the logic of life (von Neumann, 1966). In particular, von Neumann asked whether we can use purely mathematical-logical considerations to discover the specific features of biological automata that make them self-reproducing.

Von Neumann used two-dimensional CAs with 29 states per cell and a 5-cell neighborhood. He showed that a universal computer can be embedded in such cellular space, namely, a device whose computational power is equivalent to that of a universal Turing machine (Hopcroft and Ullman, 1979). He also described how a universal constructor may be built, namely, a machine capable of constructing, through the use of a "constructing arm," any configuration whose description can be stored on its input tape. This universal constructor is therefore capable, given its own description, of constructing a copy of itself, i.e., self reproduce (Figure 1.2). The terms 'machine' and 'tape' refer here to configurations, i.e., patterns of states (as defined in Section 1.2.2). The mechanisms von Neumann proposed for achieving self-reproducing structures within a cellular automaton bear strong resemblance to those employed by biological life, discovered during the following decade. Von Neumann's universal computer-constructor was simplified by Codd (1968) who used an 8-state, 5-neighbor cellular space (we shall elaborate on these issues in Chapters 2 and 3).

Over the years CAs have been applied to the study of general phenomenological aspects of the world, including communication, computation, construction, growth, reproduction, competition, and evolution (see, e.g., Burks, 1970; Smith, 1969; Toffoli and Margolus, 1987; Perrier et al., 1996). One of the most well-known CA rules, the "game of life," was conceived by Conway in the late 1960s (Gardner, 1970; Gardner, 1971) and was shown by him to be computation-universal (Berlekamp et al., 1982). For a review of computation-theoretic results, refer to Culik II et al. (1990).

The question of whether cellular automata can model not only general phenomenological aspects of our world, but also directly model the laws of physics themselves was raised by Toffoli (1977) and by Fredkin and Toffoli (1982). A primary theme of this research is the formulation of computational models of physics that are *information-preserving*, and thus retain one of the most fundamental features of microscopic physics, namely, *reversibility* (Fredkin and Toffoli, 1982; Margolus, 1984; Toffoli, 1980). This approach has been used to provide extremely simple models of common differential equations of physics, such as the heat and wave equations (Toffoli, 1984) and the Navier-Stokes equation (Hardy et al., 1976; Frisch et al., 1986). CAs also provide a useful model for a branch of dynamical systems theory which studies the emergence of well-characterized collective phenomena, such as ordering, turbulence, chaos, symmetry-breaking, and fractality (Vichniac, 1984; Bennett and Grinstein, 1985).

The systematic study of CAs in this context was pioneered by Wolfram and studied extensively by him (Wolfram, 1983; Wolfram, 1984a; Wolfram, 1984b). He investigated CAs and their relationships to dynamical systems, identifying the following four qualitative classes of CA behavior, with analogs in the field of dynamical systems (the latter are shown in parenthesis; see also Langton, 1986; Langton, 1992a):

1. Class I relaxes to a homogeneous state (limit points).

2. Class II converges to simple separated periodic structures (limit cycles).

3. Class III yields chaotic aperiodic patterns (chaotic behavior of the kind associated with strange attractors).

4. Class IV yields complex patterns of localized structures, including propagating structures (very long transients with no apparent analog in continuous dynamical systems).

Figure 1.3 demonstrates these four classes using one-dimensional CAs (as studied by Wolfram). Finally, biological modeling has also been carried out using CAs (Ermentrout and Edelstein-Keshet, 1993).

Non-uniform CAs have been investigated by Vichniac et al. (1986) who discuss a one-dimensional CA in which a cell probabilistically selects one of two rules at each time step. They showed that complex patterns appear characteristic of class IV behavior (see also Hartman and Vichniac, 1986). Garzon (1990) presents two generalizations of cellular automata, namely, discrete neural networks and

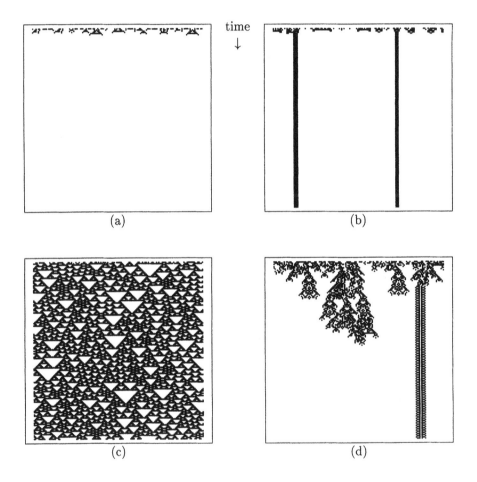

Figure 1.3. Wolfram classes. One dimensional CAs are shown, where the horizontal axis depicts the configuration at a certain time t and the vertical axis depicts successive time steps (increasing down the page). CAs are binary (2 states per cell) with radius $r = 2$ (two neighbors on both sides of the cell). (a) Class I (totalistic rule 4). (b) Class II (totalistic rule 24). (c) Class III (totalistic rule 12). (d) Class IV (totalistic rule 20).

automata networks. These are compared to the original model from a computational point of view which considers the classes of problems such models can solve. Our interest in this volume is in examining the non-uniform CA model from a computational aspect as well as an evolutionary one.

1.3   Genetic algorithms

In the 1950s and the 1960s several researchers independently studied evolutionary systems with the idea that evolution could be used as an optimization tool for engineering problems. Central to all the different methodologies is the notion of solving problems by evolving an initially random population of candidate solutions, through the application of operators inspired by natural genetics and natural selection, such that in time "fitter" (i.e., better) solutions emerge (Bäck, 1996; Michalewicz, 1996; Mitchell, 1996; Schwefel, 1995; Fogel, 1995; Koza, 1992; Goldberg, 1989; Holland, 1975). In this volume we shall concentrate on one type of evolutionary algorithms, namely, *genetic algorithms* (Holland, 1975).

Holland's original goal was not to design algorithms to solve specific problems, but rather to formally study the phenomenon of adaptation as it occurs in nature and to develop ways in which the mechanisms of natural adaptation might be imported into computer systems. Nowadays, genetic algorithms are ubiquitous, having been successfully applied to numerous problems from different domains, including optimization, automatic programming, machine learning, economics, operations research, immune systems, ecology, population genetics, studies of evolution and learning, and social systems (Mitchell, 1996). For a recent review of the current state of the art, refer to Tomassini (1996).

A genetic algorithm is an iterative procedure that consists of a constant-size population of individuals, each one represented by a finite string of symbols, known as the *genome*, encoding a possible solution in a given problem space. This space, referred to as the *search space*, comprises all possible solutions to the problem at hand. Generally speaking, the genetic algorithm is applied to spaces which are too large to be exhaustively searched. The symbol alphabet used is often binary due to certain computational advantages put forward by Holland (1975) (see also Goldberg, 1989). This has been extended in recent years to include character-based encodings, real-valued encodings, and tree representations (Michalewicz, 1996).

The standard genetic algorithm proceeds as follows: an initial population of individuals is generated at random or heuristically. Every evolutionary step, known as a *generation*, the individuals in the current population are *decoded* and *evaluated* according to some predefined quality criterion, referred to as the *fitness*, or *fitness function*. To form a new population (the next generation), individuals are *selected* according to their fitness. Many selection procedures are currently in use, one of the simplest being Holland's original *fitness-proportionate selection*, where individuals are selected with a probability proportional to their relative fitness. This ensures that the expected number of times an individual is chosen is approximately proportional to its relative performance in the population. Thus, high-fitness ("good") individuals stand a better chance of "reproducing," while low-fitness ones are more likely to disappear.

Selection alone cannot introduce any new individuals into the population, i.e., it cannot find new points in the search space; these are generated by genetically-inspired operators, of which the most well known are *crossover* and *mutation*. Crossover is performed with probability p_{cross} (the "crossover probability" or "crossover rate") between two selected individuals, called *parents*, by exchanging parts of their genomes (i.e., encodings) to form two new individuals, called *offspring*. In its simplest form, substrings are exchanged after a randomly-selected crossover point. This operator tends to enable the evolutionary process to move toward "promising" regions of the search space. The mutation operator is introduced to prevent premature convergence to local optima by randomly sampling new points in the search space. It is carried out by flipping bits at random, with some (small) probability p_{mut}. Genetic algorithms are stochastic iterative processes that are not guaranteed to converge. The termination condition may be specified as some fixed, maximal number of generations or as the attainment of an acceptable fitness level. Figure 1.4 presents the standard genetic algorithm in pseudo-code format.

> **begin GA**
>   g:=0   { generation counter }
>   Initialize population $P(g)$
>   Evaluate population $P(g)$ { i.e., compute fitness values }
>   **while** not done **do**
>   g:=g+1
>   Select $P(g)$ from $P(g-1)$
>   Crossover $P(g)$
>   Mutate $P(g)$
>   Evaluate $P(g)$
>   **end while**
> **end GA**

Figure 1.4. Pseudo-code of the standard genetic algorithm.

Let us consider the following simple example, due to Mitchell (1996), demonstrating the genetic algorithm's workings. The population consists of 4 individuals, which are binary-encoded strings (genomes) of length 8. The fitness value equals the number of ones in the bit string, with $p_{cross} = 0.7$ and $p_{mut} = 0.001$. More typical values of the population size and the genome length are in the range 50-1000. Note that fitness computation in this case is extremely simple, since no complex decoding or evaluation is necessary. The initial (randomly generated) population might look like this:

Label	Genome	Fitness
A	00000110	2
B	11101110	6
C	00100000	1
D	00110100	3

Using fitness-proportionate selection we must choose 4 individuals (two sets of parents), with probabilities proportional to their relative fitness values. In our example, suppose that the two parent pairs are {B,D} and {B,C} (note that A did not get selected as our procedure is probabilistic). Once a pair of parents is selected, crossover is effected between them with probability p_{cross}, resulting in two offspring. If no crossover is effected (with probability $1 - p_{cross}$), then the offspring are exact copies of each parent. Suppose, in our example, that crossover takes place between parents B and D at the (randomly chosen) first bit position, forming offspring E=10110100 and F=01101110, while no crossover is effected between parents B and C, forming offspring that are exact copies of B and C. Next, each offspring is subject to mutation with probability p_{mut} per bit. For example, suppose offspring E is mutated at the sixth position to form E′=10110000, offspring B is mutated at the first bit position to form B′=01101110, and offspring F and C are not mutated at all. The next generation population, created by the above operators of selection, crossover, and mutation, is therefore:

Label	Genome	Fitness
E′	10110000	3
F	01101110	5
C	00100000	1
B′	01101110	5

Note that in the new population, although the best individual with fitness 6 has been lost, the *average fitness* has increased. Iterating this procedure, the genetic algorithm will eventually find a perfect string, i.e., with maximal fitness value of 8.

The implementation of an evolutionary algorithm, an issue which usually remains in the background, is quite costly in many cases, since populations of solutions are involved, coupled with computation-intensive fitness evaluations. One possible solution is to parallelize the process, an idea which has been explored to some extent in recent years (see reviews by Tomassini, 1996; Cantú-Paz, 1995). While posing no major problems in principle, this may require judicious modifications of existing algorithms or the introduction of new ones in order to meet the constraints of a given parallel machine. The models and algorithms introduced in this volume are inherently parallel and local, lending themselves more readily to implementation. Indeed, we have already noted that one of our major goals is to attain evolware, i.e., real-world, evolving machines, an issue which shall be explored in Chapter 6.

Chapter 2

Universal Computation in Quasi-Uniform Cellular Automata

There never was in the world two opinions alike, no more than two hairs or two grains; the most universal quality is diversity.

Michael de Montaigne, *Of the Resemblance of Children to their Fathers*

2.1 Introduction

In this chapter we consider the issue of universal computation in two-dimensional CAs, namely, the construction of machines, embedded in cellular space, whose computational power is equivalent to that of a universal Turing machine (Hopcroft and Ullman, 1979). The first such machine was described by von Neumann (1966), who used 29-state, 5-neighbor cells. Codd (1968) provided a detailed description of a computer embedded in an 8-state, 5-neighbor cellular space, thus reducing the complexity of von Neumann's machine. Banks (1970) described a 2-state and a 3-state automaton (both 5-neighbor) which support universal computation with an infinite and finite initial configuration, respectively. A cellular space with a minimal number of states (two) and a 9-cell neighborhood proven to support universal computation (with a finite initial configuration) involves the "game of life" rule (Berlekamp et al., 1982). One-dimensional CAs have also been shown to support universal computation (Smith, 1992; Smith, 1971). For a review of computation-theoretic results related to this issue, refer to Culik II et al. (1990).

Codd (1968) proved that there does not exist a computation-universal, *uniform*, 2-state, 5-neighbor CA with a finite initial configuration. In this chapter we introduce computation-universal, *non-uniform* CAs, embedded in such space (Sipper, 1995b). We present three implementations, noting a tradeoff between state-space complexity (i.e., the structures of the elemental components) and rule-space complexity (i.e., the number of distinct rules). Section 2.2 presents the details of our basic system, consisting of ten different cell rules, which are reduced to six in Section 2.3. Both sections involve an infinite initial configuration. Section 2.4 describes the implementation of a universal machine using a finite initial configuration. A quasi-uniform automaton is discussed in Section 2.5, presenting a minimal implementation consisting of two rules. A discussion of our results follows in Section 2.6.

2.2   A universal 2-state, 5-neighbor non-uniform CA

In order to prove that a two-dimensional CA is computation universal we proceed along the lines of similar works and implement the following components (Berlekamp et al., 1982; Nourai and Kashef, 1975; Banks, 1970; Codd, 1968; von Neumann, 1966):[1]

1. Signals and signal pathways (wires). We must show how signals can be made to turn corners, to cross, and to fan out.

2. A functionally-complete set of logic gates. A set of operations is said to be *functionally complete* (or *universal*) if and only if every switching function can be expressed entirely by means of operations from this set (Kohavi, 1970). We shall use the *NAND* function (gate) for this purpose (this gate comprises a functionally-complete set and is used extensively in VLSI since transistor switches are inherently inverters, Millman and Grabel, 1987).

3. A clock that generates a stream of pulses at regular intervals.

4. Memory. Two types are discussed: finite and infinite.

In the following sections we describe the implementations of the above components.[2]

2.2.1   Signals and wires

A wire in our system consists of a path of connected *propagation cells*, each containing one of the *propagation rules*. A *signal* consists of a state, or succession of states, propagated along the wire. There are four propagation cell types (i.e., four different rules), one for each direction: right, left, up, and down (Table 2.1). Figure 2.1a shows a wire constructed from propagation cells. Figures 2.1b-d demonstrate signal propagation along the wire. Note that all cells which are not part of the machine (i.e., its components) contain the *NC* (No Change) rule (Table 2.1) which simply preserves its initial state indefinitely.

---

[1]Another approach is one in which a row of cells simulates the squares on a Turing machine tape while at the same time simulating the head of the Turing machine. This has been applied to one-dimensional CAs (Smith, 1992; Smith, 1971; Banks, 1970).

[2]Note- the terms 'gate' and 'cell' are used interchangeably throughout this chapter.

Designation	Rule	S	Designation	Rule	S
Right propagation cell	$\boxed{x}\;\boxed{*}\;\boxed{*}\mapsto x$ (with $\boxed{*}$ above and $\boxed{*}$ below center)	$\rightarrow$	Exclusive Or (XOR) cell (type a)	$\boxed{x}\;\boxed{*}\;\boxed{*}\mapsto x\oplus y$ (with $\boxed{y}$ above and $\boxed{*}$ below center)	$\oplus$
Left propagation cell	$\boxed{*}\;\boxed{*}\;\boxed{x}\mapsto x$ (with $\boxed{*}$ above and $\boxed{*}$ below center)	$\leftarrow$	Exclusive Or (XOR) cell (type b)	$\boxed{*}\;\boxed{*}\;\boxed{y}\mapsto x\oplus y$ (with $\boxed{x}$ above and $\boxed{*}$ below center)	$\oplus$
Up propagation cell	$\boxed{*}\;\boxed{*}\;\boxed{*}\mapsto x$ (with $\boxed{*}$ above and $\boxed{x}$ below center)	$\uparrow$	Exclusive Or (XOR) cell (type c)	$\boxed{*}\;\boxed{*}\;\boxed{x}\mapsto x\oplus y$ (with $\boxed{*}$ above and $\boxed{y}$ below center)	$\oplus$
Down propagation cell	$\boxed{*}\;\boxed{*}\;\boxed{*}\mapsto x$ (with $\boxed{x}$ above and $\boxed{*}$ below center)	$\downarrow$	Exclusive Or (XOR) cell (type d)	$\boxed{y}\;\boxed{*}\;\boxed{*}\mapsto x\oplus y$ (with $\boxed{*}$ above and $\boxed{x}$ below center)	$\oplus$
$NAND$ Cell	$\boxed{x}\;\boxed{*}\;\boxed{*}\mapsto x\mid y$ (with $\boxed{y}$ above and $\boxed{*}$ below center)	**│**	No Change (NC) cell	$\boxed{*}\;\boxed{x}\;\boxed{*}\mapsto x$ (with $\boxed{*}$ above and $\boxed{*}$ below center)	$\cdot$

Table 2.1. Cell types (rules). Rules are given in the form of "templates" rather than delineating the entire table. Each rule template specifies the transition from the current neighborhood configuration to the new state of the central cell. '*' denotes the set of states $\{0, 1\}$. $x,y \in \{0, 1\}$ denote specific states. '$\oplus$' is the XOR function (modulo-2 addition), '|' is the $NAND$ function. S is the symbol used to denote this cell type in the figures ahead. For example, the right propagation rule template specifies that when the state of the west cell is x then the state of the central cell (at the next time step) becomes x. This rule is depicted as a '$\rightarrow$' symbol in the figures ahead.

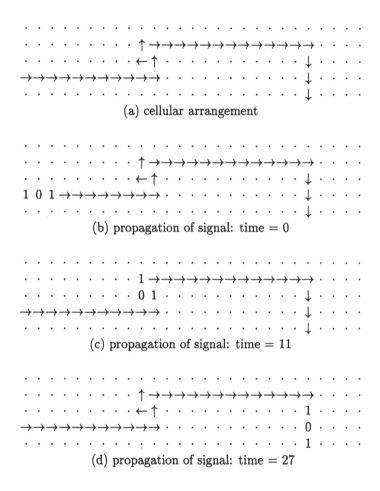

(a) cellular arrangement

(b) propagation of signal: time = 0

(c) propagation of signal: time = 11

(d) propagation of signal: time = 27

Figure 2.1. Signal propagation along a wire.

A wire in our system possesses a distinct direction, a characteristic which is highly desirable as it simplifies signal propagation (Codd, 1968). In most cases signals must propagate solely in one direction, and should bi-directional propagation be required then two parallel wires in opposite directions may be used. We note in Figure 2.1 that wires support signal propagation across corners. Fan-out of signals is also straightforward as evident in Figure 2.2.

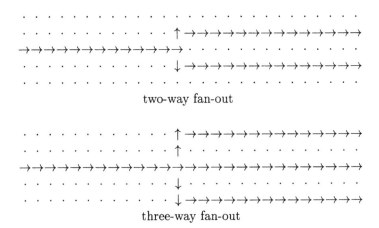

Figure 2.2. Signal fan-out.

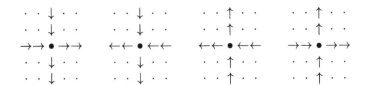

Figure 2.3. Four possible ways in which wires can cross.

The last problem we must address concerning signals is wire crossing (there are four possible crossings, see Figure 2.3). We first demonstrate that at least three gates (cells) are required for this operation. To see this note that one gate is insufficient since there are two bits of information, denoted x and y, whereas the intersection cell can only contain one bit:

$$
\begin{array}{ccccc}
\cdot & \cdot & y & \cdot & \cdot \\
\cdot & \cdot & \downarrow & \cdot & \cdot \\
x & \rightarrow & \boxed{?} & \rightarrow & \rightarrow \\
\cdot & \cdot & \downarrow & \cdot & \cdot \\
\cdot & \cdot & \downarrow & \cdot & \cdot \\
\end{array}
$$

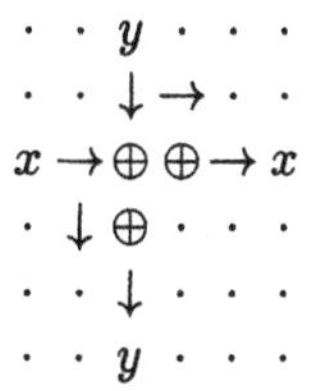

$$
\begin{array}{ccccc}
\cdot & \cdot & y & \cdot & \cdot & \cdot \\
\cdot & \cdot & \downarrow & \rightarrow & \cdot & \cdot \\
x & \rightarrow & \oplus & \oplus & \rightarrow & x \\
\cdot & \downarrow & \oplus & \cdot & \cdot & \cdot \\
\cdot & \cdot & \downarrow & \cdot & \cdot & \cdot \\
\cdot & \cdot & y & \cdot & \cdot & \cdot \\
\end{array}
$$

Figure 2.4. Implementation of wire crossing. The implementation is based on the equivalences: $x \equiv (x \oplus y) \oplus y$, and $y \equiv (x \oplus y) \oplus x$. The XOR gates used are type a (see Table 2.1).

Thus, either x crosses the y (vertical) line, in which case the y signal is lost, or, conversely, the x signal is lost. In case there are two gates then they must eventually contain x and y (otherwise information is lost). The situation is therefore as follows:

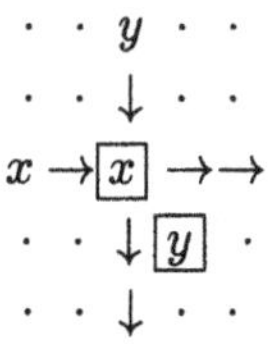

$$
\begin{array}{ccccc}
\cdot & \cdot & y & \cdot & \cdot \\
\cdot & \cdot & \downarrow & \cdot & \cdot \\
x & \rightarrow & \boxed{x} & \rightarrow & \rightarrow \\
\cdot & \cdot & \downarrow & \boxed{y} & \cdot \\
\cdot & \cdot & \downarrow & \cdot & \cdot \\
\end{array}
$$

The x signal gets transferred, however, there is no way for the y signal to get to its gate (note that the y gate must be below the x line), since this is exactly the crossing problem we are trying to solve, and we have already shown that this cannot be done using one cell (the remaining one). A similar argument holds for the reverse situation, i.e., the intersection cell contains y. We therefore conclude that at least three gates are required for the crossing operation.

Toward this end we have selected the XOR cells (Table 2.1) since wire crossing can be implemented using a minimal number of such gates (three). An implementation of one of the four possible crossings is given in Figure 2.4 (the other three are derived analogously).

Note that in uniform cellular automata the implementation of wires and signals is highly complicated (von Neumann, 1966; Codd, 1968; Banks, 1970). The wire itself and the operations of propagation, crossing, and fan-out, are attained using complex structures composed of several cells. The large number of possible interactions between these structures makes the design task arduous.

It is important to note the direct relationship between path length and time: if two paths branch out from a common point A to points B and C, and if path length AB is strictly greater than path length AC, then a signal which fans out at A will arrive at B strictly later than at C (this issue was emphasized by Codd, 1968).

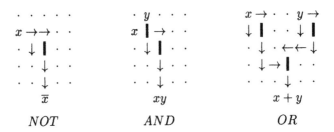

Figure 2.5. Implementations of logic gates NOT, AND, and OR, using $NAND$ gates.

2.2.2 Logic gates

Table 2.1 includes a 2-input, 1-output $NAND$ gate (cell), which forms a functionally complete set, thereby providing us with the second component discussed above. Two neighboring cells act as inputs while the central cell acts as the gate's output. As an example, Figure 2.5 shows implementations of the logic gates NOT, AND, and OR, using only $NAND$ gates.

The XOR gate is not required for completeness purposes, however, we have included it since wire crossing can be implemented using a minimal number of gates (Section 2.2.1). Four XOR cell types (rules) are needed to implement the four possible crossings of Figure 2.3. Once all crossings are possible we only need one $NAND$-gate type since the two wire inputs can always be made to arrive at the two input cells of the gate's neighborhood.[3]

2.2.3 Clock

The third component of our system is a clock that generates a stream of pulses at regular intervals. We implement this using a wire loop, i.e., a loop of propagation cells. Figure 2.6 presents the implementation. Note that any desired "waveform" can be produced by adjusting the size and contents of the loop. The implementation of the clock is facilitated due to the manner in which wires are constructed, i.e., as cellular arrangements. Thus, it is possible to obtain such a closed loop, which proves highly useful in our case.

2.2.4 Memory

A useful additional component is internal, finite memory (as opposed to the infinite, external memory discussed in Section 2.4). This is not essential to our demonstration of universal computation since the functionally-complete set of Section 2.2.2 suffices; a simple 1-bit memory unit (flip-flop) can be constructed

---

[3]Note that both signals must arrive synchronously. This is possible since delays can always be introduced by using, e.g., loops, which are feasible once crossings are implemented.

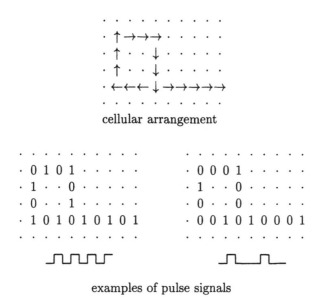

Figure 2.6. Implementation of a clock, i.e., a component that generates a stream of pulses at regular intervals. Note that any desired "waveform" can be produced by adjusting the size and contents of the loop.

from logic gates, which then serves as a basis for the construction of larger memories (Millman and Grabel, 1987). It is nonetheless interesting to note the following rule which implements a 1-bit memory unit:

The upper cell acts as a 'store' signal when set to 1, causing the bit in the left cell to be stored (left rule). After the 'store' signal is set to 0, the bit is stored indefinitely (right rule), i.e., until the storage process is repeated.

2.3   Reducing the number of rules

In the previous section we presented the components of a universal machine employing ten different cell rules (Table 2.1). This number may be reduced to six, by using a more complex wiring scheme, involving the implementation of the propagation cells using XOR gates.

Signal propagation is carried out along wires which are two cells wide. Essentially, there are two parallel paths: one of NC cells in state 0, the other consists of

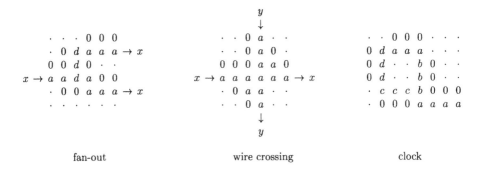

$$
\begin{array}{cccccc}
0 & 0 & 0 & 0 & 0 & 0 \\
a & a & a & a & a & a \\
 & & & & & \\
d & d & d & d & d & d \\
0 & 0 & 0 & 0 & 0 & 0
\end{array}
\qquad
\begin{array}{cccccc}
0 & 0 & 0 & 0 & 0 & 0 \\
b & b & b & b & b & b \\
 & & & & & \\
c & c & c & c & c & c \\
0 & 0 & 0 & 0 & 0 & 0
\end{array}
\qquad
\begin{array}{ccc}
0 & d & c\ 0 \\
0 & d & c\ 0 \\
0 & d & c\ 0 \\
0 & d & c\ 0 \\
0 & d & c\ 0
\end{array}
\qquad
\begin{array}{ccc}
0 & a & b\ 0 \\
0 & a & b\ 0 \\
0 & a & b\ 0 \\
0 & a & b\ 0 \\
0 & a & b\ 0
\end{array}
$$

$$\longrightarrow \qquad\qquad \longleftarrow \qquad\qquad \uparrow \qquad\qquad \downarrow$$

signal propagation (two possible implementations are shown for each direction)

fan-out   wire crossing   clock

Figure 2.7. Implementing a universal machine using a reduced number of rules (6 different rules, instead of the previous 10). a, b, c, d represent XOR gates of types a, b, c, d, respectively (Table 2.1). 0 denotes an NC cell in state 0.

XOR cells which carry the actual signal. Figure 2.7 shows the implementation of the necessary operations, signal propagation, fan-out, wire crossing, and a clock, using only four XOR cell types, the NC cell, and the $NAND$ cell, of Table 2.1.

2.4   Implementing a universal machine using a finite configuration

The components presented in the previous sections are sufficient in order to build a universal machine using an infinite initial configuration. Codd (1968) conjectured that an unbounded but boundable propagation is a necessary condition for computation universality and proved that there does not exist a uniform, 2-state, 5-neighbor universal cellular automaton with a finite initial configuration (see Section 1.2.2). Following his work, universality was implemented by using more states or larger neighborhoods (Banks, 1970; Berlekamp et al., 1982; Nourai and Kashef, 1975).

The problem with finite initial configurations involving the above components is that a computation may require an arbitrary amount of space and therefore some method must exist for increasing the information storage (memory) by arbitrarily large amounts. In order to prove universality we implement Minsky's two-register universal machine, which consists of (Minsky, 1967; Nourai and Kashef, 1975; Berlekamp et al., 1982):

1. A programming unit (finite control).

2. Two potentially infinite registers.

3. The following set of instructions:

   - Increase the contents of a register by one.
   - Decrease the contents of a register by one.
   - Test whether the contents of a register equal zero.

The finite control unit may be realized using the components described in Sections 2.2 and 2.3. The major difficulty is the implementation of the registers and the three operations associated with them. According to Codd's proof a single rule in 2-state, 5-neighbor cellular space is insufficient since unbounded but boundable propagations cannot be attained.

While other researchers have turned to cellular spaces with more states or larger neighborhoods, our approach is based on non-uniformity. We conclude from the above that the minimal number of distinct cellular rules needed to implement a register is two. Indeed, we have uncovered two such rules: one which we denote the *background* rule, the other being Banks' rule (Banks, 1970) (Figure 2.8). The implementation of a universal computer consists of a finite control unit, which occupies a finite part of the grid. All other cells contain the *background* rule except for two cellular columns, infinite in one direction, containing Banks' rule. These *register columns* originate at the upper part of the control unit and each one implements one register (Figure 2.9a); the cells in these columns are denoted register cells.

The above set of three register instructions is implemented as follows: at any given moment a register column consists of an infinite[4] number of cells in state 1, and a finite number in state 0, occupying the bottom part of the column. The number of 0s represents the register's value. Initially, both register columns (i.e., all register cells) are transformed (from state 0) to state 1, thus setting the register's value to zero. For each column, this is accomplished by setting the bottom register cell along with its left and right neighbors to 1. The two 1s on both sides act as signals which travel upward along the column, setting all its cells to 1. Figure 2.9a demonstrates this process after three time steps have taken place. Three cells have already been transformed to 1, with the fourth currently being transformed, after which the (dual) signal will continue its upward movement. The overall effect of this process is that the value of both registers is initialized to zero.

Testing whether the contents of a register equal zero is straightforward since it only involves the bottom register cell: if its state is 1, the register's value is

---

[4]More precisely, the number of cells in state 1 tends to infinity as time progresses, see ahead.

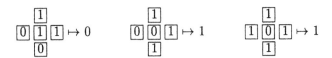

Banks' rule (note: there are three further rotations of the two left rule entries)

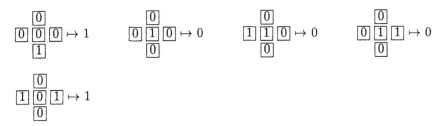

background rule

Figure 2.8. Rules used to implement registers. The figure depicts the rule tables for Banks' rule and the background rule. Only rule table entries that change the state of the central cell are shown. The other entries (not shown) preserve the state of the central cell.

zero, otherwise it is not. Adding one to a register is achieved by setting to 1 the cell which is at distance two to the right of the bottom register cell. Figure 2.9b demonstrates this operation. The left grid depicts the configuration before the operation, where the register's value is 3 and the appropriate cell is set to 1 (i.e., two cells to the right of the bottom cell). The right grid depicts the effect of the operation (i.e., the configuration after several time steps): the column's number of zeros has increased by one, which means that the register's value is now 4. Subtracting one from a register is done by setting to 1 both neighboring cells of the bottom register cell (Figure 2.9c demonstrates this operation). Thus, the registers, along with their associated instructions, have been implemented.

The initial configuration of the machine is finite, since only a finite number of cells are initially non-zero. The total number of distinct rules needed to implement a universal computer equals the number of rules necessary for implementation of the finite control unit plus the two additional memory rules (background and Banks). Thus, we need a total of 12 rules using our implementation of Section 2.2 and 8 rules using the implementation of Section 2.3.

2.5   A quasi-uniform cellular space

As noted in Section 2.3, by increasing the complexity of the basic components (in state space), a reduced set of rules (six) may be used to construct the finite

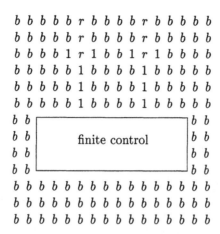

(a) initially setting both registers to zero.

```
b b b 1 b b b          b b b 1 b b b
b b b 1 b b b          b b b 1 b b b
b b b 1 b b b          b b b r b b b
b b b r b b b          b b b r b b b
b b b r b b b          b b b r b b b
b b b r b 1 b          b b b r b b b
```

(b) adding one to a register.

```
b b b 1 b b b          b b b 1 b b b
b b b 1 b b b          b b b 1 b b b
b b b 1 b b b          b b b 1 b b b
b b b r b b b          b b b 1 b b b
b b b r b b b          b b b r b b b
b b 1 r 1 b b          b b b r b b b
```

(c) subtracting one from a register.

Figure 2.9. Register operation. b denotes a cell in state 0 containing the background rule, r denotes a cell in state 0 containing Banks' rule (register). In figures (b) and (c) the left grid shows the configuration before the operation, the right grid shows the configuration upon its completion (after several time steps); the bottom line represents the bottom register cell and its neighbors.

control. We can go one step further, and construct the control unit with only one rule, e.g., Banks' rule.[5] As noted in Section 2.4, the complicating issue is not due to this unit, but rather to the infinite memory, which cannot be implemented in *uniform*, 2-state, 5-neighbor cellular space. We conclude that a universal computer can be implemented in a non-uniform cellular space with a minimal number of distinct rules (two): background and Banks.

The rules necessary to implement universal computation are distributed unevenly. Most of the grid contains the background rule, except for an infinitely small region which contains the others. By this we mean that each (infinite) row contains an *infinite* number of background rules with only a *finite* number of the others. In fact, except for a finite region of the grid, each row contains only two Banks rules and an infinite number of background rules. Hence we say that our cellular space is *quasi*-uniform.

Let n_u denote the number of rules *used* by a non-uniform CA, n_p the number of *possible* rules (i.e., the size of the rule space). Quasi-uniformity implies that $n_u \ll n_p$. We define two types of quasi-uniformity. Let $R^u = \{R_1, \ldots, R_{n_u}\}$ denote the set of rules used by the CA, and $R_j(N)$ the number of cells with rule R_j in a grid of size N, $j \in \{1, \ldots, n_u\}$. We say a grid is quasi-uniform, *type 1* if there exists $D^u \subset R^u$ such that:

$$\lim_{N \to \infty} \frac{\sum_{n \in R^u \setminus D^u} R_n(N)}{\sum_{n \in D^u} R_n(N)} = 0;$$

we say a grid is quasi-uniform, *type 2* if there exists m such that:

$$\lim_{N \to \infty} \frac{\sum_{n \neq m} R_n(N)}{R_m(N)} = 0.$$

Essentially, type 1 consists of grids in which a subset, $D^u \subset R^u$, of *dominant* rules occupies most of the grid, while type 2 consists of grids with one dominant rule, i.e., the size of set D^u is one. The computation-universal systems presented above are all quasi-uniform, type 2.

2.6   Discussion

We presented *quasi*-uniform, 2-state, 5-neighbor CAs capable of universal computation. Quasi-uniformity implies that the number of different rules used is extremely small with respect to rule-space size. Two types of quasi-uniformity were discussed: type 1 consists of grids in which a subset of dominant rules occupies most of the grid, while type 2 consists of grids with one dominant rule. We showed three type-2 universal systems using 12 rules, 8 rules, and finally 2 rules (which is minimal).

The following paragraphs provide a discussion of a speculative nature, linking our results with those of Langton (1992a) (see also Li et al., 1990). He addressed

---

[5]This increases the complexity of the basic operations. As noted above, there is a tradeoff between state-space complexity and rule-space complexity.

the following question: under what conditions can we expect a dynamics of information to emerge spontaneously and come to dominate the behavior of a physical system? This was studied in the context of CAs where the question becomes: under what conditions can we expect a complex dynamics of information to emerge spontaneously and come to dominate the behavior of a CA? (Langton, 1992a).

Langton showed that the rule space of (uniform) CAs consists of two primary regimes of rules, periodic and chaotic, separated by a transition regime. His main conclusion was that information processing can emerge spontaneously and come to dominate the dynamics of a physical system in the vicinity of a critical phase transition (see also Section 4.2).

According to Codd's proof a uniform (single rule), 2-state, 5-neighbor cellular space is insufficient for universal computation since unbounded but boundable propagations cannot be attained; either every configuration yields an unboundable propagation or every configuration yields a bounded propagation (Codd, 1968). In the context of Langton's work, bounded propagations correspond to fixed-point rules (class I) and unboundable propagations correspond either to periodic rules (class II) or chaotic ones (class III). Complex behavior (class IV) cannot be attained.[6]

By using a quasi-uniform, two-rule cellular space we have been able to achieve unbounded but boundable propagations, thus attaining class-IV behavior. Langton suggested that the information dynamics which gave rise to life came into existence when global or local conditions brought some medium through a critical phase transition.

Imagine an information-based world, consisting of a uniform cellular automaton, which is not within the class-IV region. If we wanted to attain class-IV behavior, the *entire* space would have to "jump," i.e., undergo a phase transition, to a class-IV rule (assuming this is at all possible– e.g., in the case of a 2-state, 5-neighbor space it is not). However, as a conclusion of the work presented above we offer an alternative: a small perturbation may be enough to cause some infinitely small part of the world to change. This would be sufficient to induce a (possible) transition from class-II or class-III behavior to class-IV behavior. Furthermore, such a change could be effected upon a very simple world (in our case, 2-state, 5-neighbor). As noted by Bonabeau and Theraulaz (1994), "frozen accidents" play an important role in the evolutionary process. These accidents are mainly caused by external conditions, i.e., external relative to a given system's laws of functioning.

A (highly) tentative comparison may be drawn with the famous experiment performed by Miller (1953) (see also Miller and Urey, 1959), in which methane, ammonia, water, and hydrogen, representing a possible atmosphere of the primitive Earth, were subjected to an electric spark for a week. After this period simple amino acids were found in the system. The analogy to our CA world is as follows: we start with a simple uniform world, consisting of a single rule,

---

[6]Class numbers are those defined by Wolfram, see Section 1.2.4.

which does not support complex (class-IV) behavior.[7] At some point, a "spark" causes a perturbation in which a small number of cells change their rule. Such an infinitely small change in our world can suffice to generate a phase transition such that class-IV behavior becomes possible. Note that this can happen independently in other regions of the world as well.

While the above discussion has been of a tentative, speculative nature, we may also draw some practical conclusions from this chapter. As noted in Section 1.2.3, a primary difficulty with the CA approach lies with the extreme low-level representation of the interactions; essentially, we construct world models at the level of "physics." By slightly changing the rules of the game (no pun intended) we can increase the "capacity" for complex computation and ALife modeling, while preserving the main features of CAs, namely, massive parallelism, locality of cellular interactions, and simplicity of cells. After demonstrating that simple, non-uniform CAs comprise viable parallel cellular machines, we proceed in the next chapter to study ALife issues in such a model. In the succeeding chapters we present the *cellular programming* approach, by which parallel cellular machines are evolved to perform computational tasks.

---

[7]either due to inability of the cellular space to support such behavior at all or due to the rule being in the non-class-IV regions of rule space.

Chapter 3

Studying Artificial Life Using a Simple, General Cellular Model

> *Four things there are which are smallest on earth*
> *yet wise beyond the wisest:*
> *ants, a people with no strength,*
> *yet they prepare their store of food in the summer;*
> *rock-badgers, a feeble folk,*
> *yet they make their home among the rocks;*
> *locusts, which have no king,*
> *yet they all sally forth in detachments;*
> *the lizard, which can be grasped in the hand,*
> *yet is found in the palaces of kings.*
>
> Proverbs 30, 24-28

3.1 Introduction

A major theme in the field of artificial life (ALife) is the emergence of complex behavior from the interactions of simple elements. Natural life emerges out of the organized interactions of a great number of non-living molecules, with no global controller responsible for the behavior of every part (Langton, 1989). Closely related to the concept of emergence is that of evolution, in natural settings, as well as in artificial ones.

Several major outstanding problems in biology are related to these two themes, emergence and evolution, among them (Taylor and Jefferson, 1994): (1) How do populations of organisms traverse their adaptive landscapes- through gradual fine-tuning by natural selection on large populations, or alternatively in fits and starts with a good bit of chance to "jump" adaptive valleys in order to find more favorable epistatic combinations? (2) What is the relation between adaptedness and fitness, that is, between adaptation and what is selected for? It is now understood that natural selection does not necessarily maximize adaptedness, even in theory (Mueller and Feldman, 1988). Factors such as chance, structural necessity, pleiotropy, and historical accident, detract from the "optimization in nature" argument (Gould and Lewontin, 1979; Kauffman, 1993). (3) The formation of multicellular organisms from basic units or cells. Other problems include the origin of life, cultural evolution, the origin and maintenance of sex, and the structure of ecosystems (Taylor and Jefferson, 1994).

This is just a partial list of open problems amenable to study by ALife modeling. ALife research into such issues holds a potential two-fold benefit: (1) increasing our understanding of biological phenomena, and (2) enhancing our understanding of artificial models, thereby providing us with the ability to improve their performance (e.g., robotics, evolving software).

Our main interest in this chapter lies in studying evolution, adaptation, and multicellularity, in a model which is both *general* and *simple*. Generality implies two things: (1) the model supports universal computation, and (2) the basic units encode a general form of local interaction rather than some specialized action (e.g., an IPD strategy, see Section 3.4.3). Simplicity implies that the basic units of interaction are "modest" in comparison to Turing machines. If we imagine a scale of complexity, with Turing machines occupying the high end, then simple machines are those that occupy the low end, e.g., finite state automatons. These two guidelines, generality and simplicity, allow us to evolve complex behavior with the ability to explore, in-depth, the inner workings of the evolutionary process (we shall come back to this point in the discussion in Section 3.5).

The CA model is perhaps the simplest, general model available. The basic units (cells) are simple, local, finite state machines, representing a general form of local interaction; furthermore, CAs support universal computation (Chapter 2). As noted in Chapter 1, the main difficulty with the CA approach seems to lie with the extreme low-level representation of the interactions. CAs are programmed at the level of the local physics of the system and therefore higher-level cooperative structures are difficult to evolve (Rasmussen et al., 1992). Our intent is to increase the "capacity" for ALife modeling, while preserving the essential features of the CA model, namely, massive parallelism, locality of cellular interactions, and simplicity of cells.

The ALife model studied in this chapter is detailed in Section 3.2 and the evolutionary aspect is presented in Section 3.4.1. The three basic features by which it differs from the original CA model are (Sipper, 1994; Sipper, 1995c):

1. Whereas the CA model consists of uniform cells, each containing the same rule, we consider the non-uniform case where different cells may contain different rules.

2. The rules are slightly more complex than CA rules.

3. Evolution takes place not only in *state space* as in the CA model, but also in *rule space*, i.e., rules may change (evolve) over time.

Thus, we obtain a grid of simple, interacting, rule-driven "organisms" that evolve over time. The course of evolution is influenced by the nature of these organisms, as well as by their *environment*. In nature, the role of the environment in generating complex behavior is well known, e.g., as noted by Simon (1969) who described a scene in which the observed complexity of an ant's path is due to the complexity of the environment and not necessarily a reflection of the complexity of the ant. In our model, each rule is considered to have a certain *fitness*, depending upon the environment under consideration. As opposed to the standard genetic algorithm (Section 1.3), where each individual in the population is independent, interacting only with the fitness function (and not the environment), in our case fitness depends on *interactions* of evolving organisms, operating in an environment (see also Section 3.4.1).

Note that the term 'environment' can convey two meanings: in the strict sense it refers to the surroundings, excluding the organisms themselves (e.g., sun, water, and the climate, in a natural setting), while the broad sense refers to the total system, i.e., surroundings + interacting organisms (e.g., ecosystem). In what follows the term is used in the strict sense, however, we attain an environment in the broad sense, i.e., a total system of interacting organisms (see also Bonabeau and Theraulaz, 1994).

We consider various environments, including the basic environment where rules compete for space on the grid, an IPD (Iterated Prisoner's Dilemma) environment, an environment of spatial niches, and an environment of temporal niches. One of the advantages of ALife models is the opportunities they offer in performing in-depth studies of the evolutionary process. This is accomplished in our case by observing not only phenotypic effects (i.e., cellular states as a function of time) but also such measures as fitness, operability, energy, and the genescape.

Our approach in this chapter is an ALife one where cellular automata provide us with "logical universes" (Langton, 1986). These are "synthetic universes defined by simple rules ... One can actually construct them, and watch them evolve." (Toffoli and Margolus, 1987)

In the next section we detail the basic model (without evolution which is presented in Section 3.4.1). In Section 3.3 we present designed multicellular organisms which display several behaviors, including reproduction, growth, and mobility. These are interesting in and of themselves and also serve as motivation for the following section (Section 3.4) in which we turn our attention from the human watchmaker to the blind one, focusing on evolution (Dawkins, 1986). A discussion of our results ensues in Section 3.5.

3.2 The ALife model

The two-dimensional CA model consists of a two-dimensional grid of cells, each containing the same rule, according to which cell states are updated in a synchronous, local manner (Section 1.2). The model studied in this chapter consists of a grid of cells which are either *vacant*, containing no rule, or *operational* containing a finite state automaton (rule) which can, in one time step:

1. Access its own state and that of its immediate neighbors (grid is toroidal).

2. Change its state and the states of its immediate neighbors. Contention occurs when more than one operational neighbor attempts to change the state of the same cell. Such a situation is resolved randomly, i.e., one of the contending neighbors "wins" and decides the cell's state at the next time step. Note that the cell itself is also a contender, provided it is operational.

3. Copy its rule into a neighoring *vacant* cell. Contention occurs if more than one operational neighbor attempts to copy itself into the same cell. Such a situation is resolved randomly, i.e., one of the contending neighbors "wins" and copies its rule into the cell. Note that in this case the cell itself is not a contender since it must be vacant in the first place for contention to occur.

At each time step every operational rule[1] simultaneously executes its appropriate rule entry, i.e., the entry corresponding to its current neighborhood configuration. Thus, state changes and rule copies are effected as explained above. Our extended rule may be readily encoded in the form of a table as with the original CA rule. Figure 3.12 depicts such an encoding for a 2-state, 5-neighbor cellular space. Note that a vacant cell may be in any grid state as it can be changed by neighboring operational cells.

Whereas a cell in the CA model accesses the states of its neighbors but may only change its own state, our model allows state changes of neighboring cells and rule copying into them. Thus, our rules may be regarded as being more "active" than those of the CA model; furthermore, different cells may contain different rules (non-uniformity). The third feature of our model, as presented in Section 3.1, is the evolution that takes place in rule space, i.e., rules evolve as time progresses; this is detailed in Section 3.4.1.

Our model is essentially non-deterministic since contention is resolved randomly. A deterministic version could be attained by specifying, e.g., that upon contention no change occurs (i.e., no state change or rule copy is effected). The issue of determinism versus non-determinism has a long history whose discussion is beyond our scope. The question has been raised specifically in relation to artificial life, where it has been argued that computer experiments, by their deterministic nature, can never attain the characteristics of true living systems, where randomness is of crucial importance. However, in contrast to this objection we note that random events are indeed incorporated into ALife systems. In fact, von Neumann himself proposed, although he did not have the chance to design, a probabilistic version of his self-reproducing CA, which obviated the deterministic nature of his previous version (Section 1.2.4). For a discussion of the issue of determinism in artificial life see Levy (1992), pages 337-338.[2]

---

[1]Throughout this chapter we use the terms "operational cell" and "operational rule" interchangeably.

[2]Note that our self-reproducing loop, presented in Section 3.3.1, is in fact deterministic since contention does not arise.

What can be said about the "power" of our model in relation to the original CA? This question must be considered with some care. In terms of computational power, we have seen in Chapter 2 that non-uniformity does indeed engender a basic difference for very simple cellular spaces (we shall also see this in Chapter 4). For example, 2-state, 5-neighbor uniform CAs are not computation universal whereas non-uniform CAs are (Chapter 2). For most uniform cellular spaces universal computation can, however, be attained, so in this respect we have not increased the power of our system. In fact, it is easy to see that a uniform CA can simulate a non-uniform one by encoding all the different rules as one (huge) rule, employing a large number of states. Another feature of our model, namely, the "active" nature of our rules, whereby they may effect changes upon neighboring cells, may also be obviated by using "static" rules with larger neighborhoods, performing the equivalent operations. While the above arguments hold true in *principle* we argue that this is not so in *practice*.[3] The power offered by our model cannot strictly be reduced to the question of computational power. As noted in Section 3.1, our intent is to increase the "capacity" for ALife modeling. This notion cannot be precisely defined at this point and is in fact one of the major concerns of the field of artificial life. Nonetheless, our investigations reported in the following sections do indeed show that our model holds potential for the exploration of ALife phenomena.

3.3   Multicellularity

In this section we present a number of multicellular organisms which are composed of several cells, consisting of rules as described in Section 3.2. The organisms discussed below are designed rather than evolved and our intent is to demonstrate that interesting behaviors can arise using the dynamics described above. In the next section we shall focus on evolution. At this point the term 'multicellular' is loosely defined so as to refer to any structure composed of several cells, acting in unison. In Section 3.5, we examine more carefully the meaning of the term 'cell,' and expand upon the general issue of multicellular organisms versus unicellular ones. The cellular space considered throughout this section is 3-state, 9-neighbor, where states are denoted $\{0, 1, b\}$.[4]

3.3.1   A self-reproducing loop

Our first example involves a simple self-reproducing loop, motivated by Langton's work, who described such a structure in uniform cellular automata (Langton, 1984; Langton, 1986). His loop was later simplified by Byl (1989) and by Reggia et al. (1993). Langton's loop (motivated by Codd, 1968) makes dual use of the

---

[3]In his book on complex systems, Pagels discusses this issue in the context of Kant's epistemic dualism, which suggests that there are two different kinds of reason, "theoretical reason" (in principle) and "practical reason" (in practice) (Pagels, 1989, pages 216-222).

[4]The third state is denoted b rather than 2 since it is depicted as a blank square in the figures.

information contained in a description to reproduce itself. The structure consists of a looped pathway, containing instructions, with a construction arm projecting out from it. Upon encountering the arm junction, the instruction is replicated, with one copy propagating back around the loop again and the other copy propagating down the construction arm, where it is translated as an instruction when it reaches the end of the arm (Figure 3.1).

```
 . . . . . . . .           . . . . . . . .       . . . . . . . .
 . 1 7 0 1 4 0 1 4 .       . 7 0 1 7 0 1 7 0 .    . 1 4 0 1 4 0 1 1 .
 . 0 . . . . . . 0 .       . 1 . . . . . . 1 .    . 0 . . . . . . 1 .
 . 7 .         . 1 .       . 1 .         . 7 .    . 7 .         . 1 .
 . 1 .         . 1 .       . 1 .         . 0 .    . 1 .         . 1 .
 . 0 .         . 1 .       . 1 .         . 1 .    . 0 .         . 7 .
 . 7 .         . 1 .       . 1 .         . 7 .    . 7 .         . 0 .
 . 1 . . . . . . 1 . . . . . . 0 . . . . . . 0 . . . 1 . . . . . . 1 .
 . 0 7 1 0 7 1 0 7 1 1 1 1 1 . . 4 1 0 4 1 0 7 1 0 7 5 0 6 1 0 7 1 0 7 .
 . . . . . . . . . . . . . .   . . . . . . . . . . . . . . . . . . . .
          time = 0                              time = 126
```

Figure 3.1. Langton's self-reproducing loop.

The important issue to note is the two different uses of information, interpreted and uninterpreted, which also occur in natural self-reproduction, the former being the process of *translation*, and the latter *transcription*. In Langton's loop translation is accomplished when the instruction signals are "executed" as they reach the end of the construction arm, and upon the collision of signals with other signals. Transcription is accomplished by the duplication of signals at the arm junctions (Langton, 1984).

The loop considered in this section consists of five cells and reproduces within six time steps . The initial configuration consists of a grid of vacant cells (i.e., with no rule) with a single loop composed of five cells in state 1, each containing the loop rule (Figure 3.2a). The arm extends itself by copying its rule into an adjoining cell, coupled with a state change to that cell. The new configuration then acts as data to the arm, thereby providing the description by which the loop form is replicated. When a loop finds itself blocked by other loops it "dies" by retracting the construction arm. Figure 3.2b shows the configuration after several time steps.

The loop rule is given in Figure 3.3. Note that most entries are identity transformations, i.e., they transform a state to itself, thereby causing no change (only 40 entries of the 3^9 are non-identity). In his paper, Langton (1984) compares the self-reproducing loop with the works of von Neumann (1966) and Codd (1968), drawing the conclusion that although the capacity for universal construction, presented by both, is a *sufficient* condition for self-reproduction, it is not a *necessary* one. Furthermore, as Langton points out, naturally self-reproducing systems are not capable of universal construction. His intent was therefore to present a sim-

```
11                 11                 11                 11
111                1110               11100              11101
                                                             1
time = 0           time = 1           time = 2           time = 3


                                      1
11                 11                 11
11101              11011              11 11
   11                 11                 11
                       1                  1
                                          0
time = 4           time = 5           time = 6
```

(a)

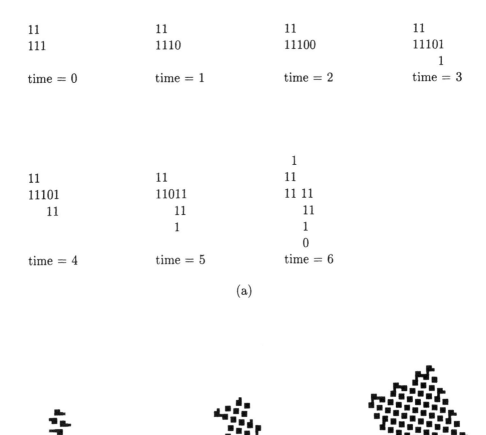

time = 12   time = 28   time = 66

(b)

Figure 3.2. Self-reproducing loop. In (b), black squares represent cells in state 1, non-filled squares represent cells in state 0, and white squares represents cells in state *b*.

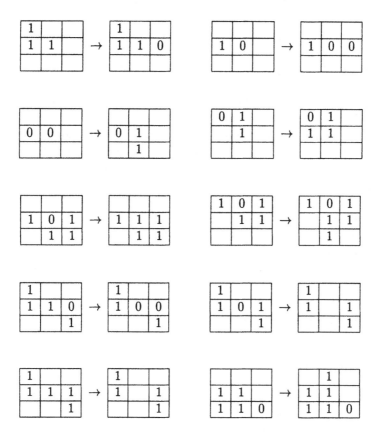

Figure 3.3. Self-reproducing loop: Rule table. In all rule entries a state change from b to 0/1 also involves a rule copy (note that all cells are initially vacant, i.e., with no rule, except the ones comprising the initial loop). For example, the upper left rule entry specifies a state change from b to 0 to the east cell, along with a rule copy to that cell. Each of the above entries consists of three further rotations (not shown). All other entries preserve the configuration.

pler system that exhibits non-trivial self-reproduction. This was accomplished by constructing a rule in an eight-state cellular space, that exhibits the dual nature of information, i.e., translation and transcription.

In the loop presented above, simple transcription is accomplished as an integral part of a cell's operation, since a rule can be copied, i.e., treated as data. Once a rule is activated it begins to function by changing states in accordance with the grid configuration, thereby performing translation on the surrounding cells (data). Essentially, the loop operates by transcribing itself onto a neighboring cell while simultaneously writing instructions (in the form of grid states) that will be carried out at the next time step.

In Langton's system each grid cell initially contains the rule that supports replication whereas in our case the grid cells are initially vacant and the loop itself contains all the information needed. In both cases reproduction is not coded entirely into the "transition physics" but rather is "actively directed by the configuration itself" where "the structure may take advantage of certain properties of the transition function physics of the cellular space" (Langton, 1984). Thus, interest in such systems arises since they display an interplay of active structures taking advantage of the characteristics of cellular space.

Before ending this section, we mention the recent work of Perrier et al. (1996) who observed that self-reproducing, cellular automata-based systems developed to date broadly fall under two categories, essentially representing two extremes. The first consists of machines which are capable of performing elaborate tasks, yet are too complex to simulate (e.g., von Neumann, 1966; Codd, 1968), while the second consists of simple machines which can be entirely implemented, yet are only capable of self-reproduction (e.g., Langton, 1984; Byl, 1989; Reggia et al., 1993, and the system described above). An interesting system situated in the middle ground was presented by Tempesti (1995). Essentially a self-reproducing loop, similar to that of Langton's, it has the added capability of attaching to the automaton an executable program which is duplicated and executed in each of its copies. The program is stored within the loop, interlaced with the reproduction code, and is therefore somewhat limited. Perrier et al. (1996) demonstrated a self-reproducing loop that is capable of implementing any program, written in a simple yet universal programming language. The system consists of three parts, loop, program, and data, all of which are reproduced, followed by the program's execution on the given data. This system has been simulated *in its entirety*, thus attaining a viable, self-reproducing machine with programmable capabilities (Figure 3.4).

3.3.2 Reproduction by copier cells

In the previous section we described a self-reproducing loop, which exhibited a two-fold utilization of information, i.e., translation and transcription. In this section we examine a system of reproduction consisting of passive structures copied by active (mobile) cells. The motivation for our approach lies in the information flow in protein synthesis, where passive mRNA structures are translated into

```
          . . . . . . . . .
          . 7 0 1 7 0 1 7 0 .
          . 1 . . . . . . 1 .
          . 1 .         . 7 .
          . 1 .         . 0 .
          . 1 .         . 1 .
          . 1 .         . 7 .
          . 0 . . . . . . 0 . . . .
          . 4 1 0 4 1 0 7 1 0 7 1 1 .
          . A . . . . . . . . . . . .
          . P .         . D .
          . P .         . D .
          . P .         . D .
          . P .         . D .
          . P .         .
          . P .
          . P .
          .
```

Figure 3.4. A self-reproducing loop with programmable capabilities. The system consists of three parts, loop, program, and data, all of which are reproduced, followed by the program's execution on the given data. P denotes a state belonging to the set of program states, D denotes a state belonging to the set of data states, and A is a state which indicates the position of the program.

amino acids by active tRNA cells. Each tRNA cell matches one specific codon in the mRNA structure and synthesizes one amino acid. Note that our system is extremely simple with regards to the workings of the living cell and therefore the above analogy is (highly) abstracted.

Our system consists of stationary structures, composed of vacant grid cells, comprising the passive data to be copied. The copy ("synthesis") process is effected by three types of copier cells, denoted X, Y, and Z, which are mobile units, "swimming" on the grid, seeking an appropriate match (remember that cellular mobility is possible by using rule copying, see Section 3.2). When such a match occurs the cell proceeds to create the appropriate sub-structure, as in the case of a tRNA cell synthesizing the appropriate amino acid. The final result is a copy of the original structure.

The process is demonstrated in Figure 3.5. The initial configuration consists of a passive structure with X, Y, and Z cells randomly distributed on the grid (Figure 3.5, $time = 0$). Each time step the copier cells move to a neighboring vacant cell (shown as white squares) at random, unless a match is found which triggers the synthesis process. Figure 3.5 shows the process at an intermediate stage ($time = 435$), and at the final stage ($time = 813$) when the copy has been produced.

The X-cell rule table is detailed in Figure 3.6 (Y- and Z-cell rules may be analogously derived). The table entry shown at the top left is the match seeker, specifying the "codon" of the X cell. Once a match is found, the cell builds a copy by applying the other two entries. After application of the entry at the bottom left, the copy has been constructed and the X cell "dies." Note that

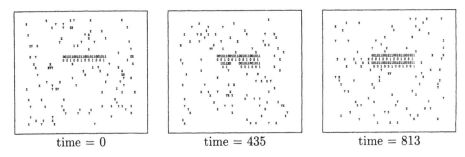

time = 0   time = 435   time = 813

Figure 3.5. Reproduction by copier cells.

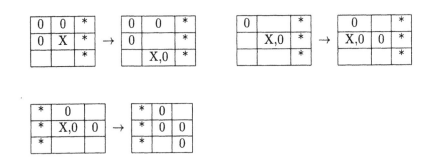

Figure 3.6. Reproduction by copier cells: X-cell rule table. Rather than provide an exhaustive listing of all table entries, it is given in the form of entry "templates" (as in Table 2.1), using the symbol '*' to denote the set of states $\{0, 1, b\}$. 'X' denotes an X rule in a cell in state b, '$X, 0$' denotes an X rule in a cell in state 0. All other entries specify a move to a random vacant cell in state b.

most entries in the rule table specify a move to a random vacant cell in state b.

The copy created is not an exact duplicate but rather a "complementary" one. The reason for this is that we wish to avoid endless copying which would occur had an exact duplicate been created. Since our model is inherently local we cannot maintain a global variable specifying that the synthesis process has been completed. The only way to avoid an endless chain of duplicate sub-structures is by locally specifying that a copy has been completed. This is accomplished by creating a complementary sub-structure, which does not match any copier cell and is not further duplicated.

3.3.3 Mobility

In this section we introduce a worm-like structure which has the capacity to move freely on the grid. The system consists of "worms," which are active, mobile structures composed of operational cells in state 1, and barriers, which are vacant cells in state 0. When a worm encounters a barrier it turns by 90 degrees and continues its movement (if there is a barrier obstructing the turn then the worm destroys it).

Figure 3.7 presents a system with a single worm, behaving as described above. When several worms are placed on the grid, interactions among them yield interesting phenomena (Figure 3.8). The following behavioral patterns are observed when two worms meet: one of them splits into two, both worms merge into one, a worm loses part of its body, or both emerge unscathed. In all cases the resulting worms behave in the same manner as their "ancestors."

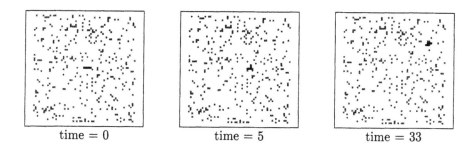

Figure 3.7. A system consisting of a single worm. Black squares represent cells in state 1 (worms), non-filled squares represent cells in state 0 (barriers), and white squares represent cells in state b.

The rule is detailed in Figure 3.9. Its simplicity is possible due to the power offered by our model (see discussion in Section 3.5). The emergent behavior is complex and exhibits different forms of interaction between the organisms inhabiting the grid. A worm acts as a single, high-order structure and upon encountering other worms it may split, merge, shrink, or emerge unscathed.

It is interesting to observe the formation of such a high-order structure which operates by applying local rules. The worm rule essentially specifies how the head and tail sections operate independently, the overall effect being that of a single

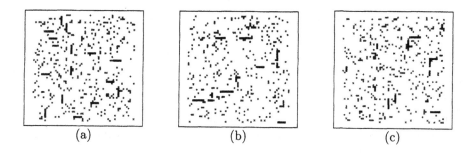

Figure 3.8. A system consisting of several worms. (a) an initial configuration of the system. (b), (c) system configurations after several time steps.

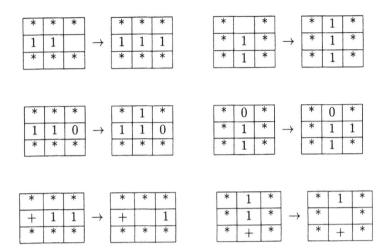

Figure 3.9. Worm rule. '*' denotes the set of states: $\{0, 1, b\}$. '+' denotes the set of states: $\{0, b\}$. All other entries preserve the configuration.

organism whose parts operate in unison. Living creatures may also be viewed in this manner, i.e., as a collection of independent cells operating in unison, thereby achieving the effect of a single "purposeful" organism (see discussion in Section 3.5).

3.3.4 Growth and replication

In this section we examine an enhancement of our model, in which the following feature is added to the three presented in Section 3.2:

4. A cell may contain a *small* number of different rules. At a given moment only one rule is *active* and determines the cell's function. An inactive rule may be activated or copied into a neighoring cell.

This feature could serve as a possible future enhancement in the evolutionary studies as well (Section 3.4). At this point we present a system involving the growth and replication of complex structures which are created from grid cells and behave as multicellular organisms once formed. The system consists initially of two cell types, builders (A cells) and replicators (B cells), floating around on the grid.

Figure 3.10 demonstrates the operation of the system. At time 0, A and B cells are distributed randomly on the grid and there are two vacant cells in state 1, acting as the core of the building process. The A cells act as builders by attaching ones at both ends of the growing structure. Once a B cell arrives at an end, growth stops there by attaching a zero (time=111).

When a B cell arrives at the upper end of a structure already possessing one zero, a C cell is *spawned*, which travels down the length of the structure to the other end. If that end is as yet uncompleted, the C cell simply waits for its completion (time=172). The C cell then moves up the structure, duplicating its right half which is also moved one cell to the right (time=179). Once the C cell reaches the upper end it travels down the structure, spawns a D cell at the bottom and begins traveling upward, while duplicating and moving the right half (time=187). Meanwhile, the D cell travels upward between two halves of the structure and joins them together (time=190).

This process is then repeated. The C cell travels up and down the right side of the structure, creating a duplicate half on its way up. As it reaches the bottom end, a D cell is spawned, which travels upward between two disjoint halves and joins them together. Since joining two halves only occurs every second pass, the D cell immediately dies every other pass (e.g., time=195).

There are interesting features to be noted in the process presented. Replication should begin only after the organism is completely formed, i.e., there are two distinct phases of development; however, there can be no global indicator that such a situation has occurred (as noted in Section 3.3.2). Our solution is therefore local: a B cell, upon encountering an upper end which already has one zero, completes the formation of that end and releases a C cell, which travels down the length of the structure. This cell will seek the bottom end or *wait* for its completion. Only at such time when the structure is complete will the C cell begin the replication process.

Replication involves two cells operating in unison, where the C cell duplicates half of the structure, while the D cell "glues" two halves together. Again, it is crucial that the whole process be local in nature since no global indicators can be used.

The rules involved in the system are given in Appendix A. The spawning of C and D cells are provided for by the added feature above, which specifies that a cell may contain a small number of different rules, where only one is active at a given moment. Therefore, the initial B cells can contain all three rules: B,C,D.

The design of our system is even more efficient than that, however, requiring only two rule tables, one for A cells and one for $B/C/D$ cells. Each entry of the $B/C/D$ rule table is only used by one of the cell types (i.e., the entries are mutually exclusive). At a given moment, the cell has one active rule (which determines its type). If the table entry to be accessed belongs to the active rule it is used, otherwise, a default state change occurs; this default transformation is a move to a random vacant cell for B cells and no change for C and D cells.

3.4 Evolution

3.4.1 Evolution in rule space

The previous section presented a number of designed, multicellular organisms, using the model delineated in Section 3.2. These organisms demonstrate the capability of our model in creating systems of interest, which results from increasing

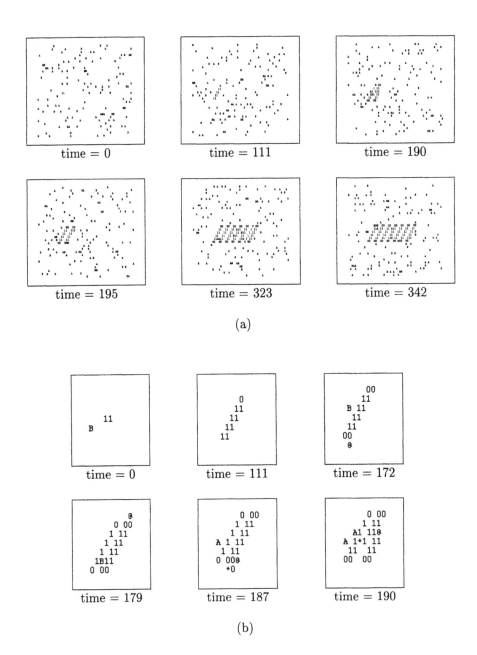

Figure 3.10. Growth and replication. Four cell types, denoted A, B, C, and D, interact to "grow" a structure, starting from a core of two vacant cells in state 1. Upon termination of the growth process, the complete structure is then replicated. (a) Overview of the entire process. (b) Zoom of intermediate stages (with C cells represented by @, and D cells by *).

the level of operation with respect to the "physics" level of CAs (Section 3.1). In this section we study evolution as it occurs in our model. Though at this point we have not yet evolved organisms as complex as those of the previous section, we have, nonetheless, encountered several interesting phenomena. We shall also present various tools with which the evolutionary process can be investigated.

Figure 3.11. The 5-cell neighborhood.

The cellular space considered in this section is 2 state, 5-neighbor (Figure 3.11), where states are denoted $\{0, 1\}$. We chose this space due to practical considerations, as well as the desire to study the simplest possible two-dimensional space. Evolution in rule space is achieved by constructing the *genome* of each cell, specifying its rule table, as depicted in Figure 3.12. There are 32 genes corresponding to all possible neighborhood configurations. Each gene consists of 10 bits, encoding the state change to be effected on neighboring cells (including itself), and whether the rule should be copied to neighboring cells or not (including itself). When discussing specific genes we will use the following notation:

$$CNESW \Rightarrow Z_C Z_N Z_E Z_S Z_W,$$

where $CNESW$ represents a neighborhood configuration, and $Z_C Z_N Z_E Z_S Z_W$ represents the respective S_x and C_x bits, using the following notation for Z_x:

	$C_x = 0$	$C_x = 1$
$S_x = 0$	$Z_x =$ '0'	$Z_x =$ '−'
$S_x = 1$	$Z_x =$ '1'	$Z_x =$ '+'

For example, $00101 \Rightarrow 01 + +-$, means that the gene (entry) 00101 specifies the following transformations: $S_C = 0, C_C = 0, S_N = 1, C_N = 0, S_E = 1, C_E = 1, S_S = 1, C_S = 1, S_W = 0, C_W = 1$.

At each time step, every operational rule simultaneously executes its appropriate rule entry by referring to the gene corresponding to its current neighborhood states, i.e., state changes and rule copies are effected as delineated in Section 3.2. This is followed by application of the "genetic" operators of crossover and mutation, as used in the standard genetic algorithm (Section 1.3).

Crossover is performed in the following manner: at each time step, every operational cell selects an operational neighbor at random. Let (i, j) denote the grid position of an operational cell and (i_n, j_n) the grid position of the randomly selected operational neighbor. Crossover is performed between the genomes of the

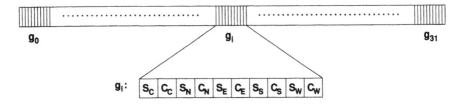

g_i - gene i corresponds to neighborhood configuration i, where i equals the binary representation of the neighboring cell states in the order $CNESW$.

S_x - state-change bit, where $x \in \{C, N, E, S, W\}$ denotes one of the five neighbors. This bit specifies the state change to be effected upon the appropriate neighboring cell. For example, $S_E = 0$ means "change the east cell's state to 0."

C_x - copy-rule bit, where $x \in \{C, N, E, S, W\}$ denotes one of the five neighbors. This bit specifies whether to copy the rule to the cell in direction x or not (0 - don't copy, 1 - copy).

Figure 3.12. Rule genome.

rules in cell (i, j) and cell (i_n, j_n), with probability p_{cross}. The (single) crossover site is selected with uniform probability over the entire string and the resulting genome is placed in cell (i, j). If the cell has no operational neighbors then no crossover is effected. Note that the crossover operator is somewhat different than the one used in genetic algorithms, due to its "asymmetry:" cell (i, j) selects cell (i_n, j_n), while cell (i_n, j_n) may select a different cell, i.e., cell (i', j') such that $(i', j') \neq (i, j)$. It is argued that this slightly decreases the coupling between cells, thus enhancing locality and generality.

Mutation is applied to the genome of each operational rule, after the crossover stage, by inverting each bit with probability p_{mut}. Note that both operations are insensitive to gene boundaries, which is also the case in biological settings. In summary, at each time step every operational rule performs its appropriate action, after which crossover and mutation are applied.

It is important to note the difference between our approach and genetic algorithms. Though we apply genetic operators in a similar fashion, there is no selection mechanism operating on a global level, using the total fitness of the entire population. As we shall see (Section 3.4.3) fitness will be introduced, albeit in a local manner consistent with our model (see also Collins and Jefferson, 1992). Note also that in the standard genetic algorithm each entity is an independent coding of a problem solution, interacting only with the fitness function, "seeing" neither the other entities in the population nor the general environment that exists (see also Ray, 1994a). In contrast, in our case fitness depends on interactions

General parameters	time steps	3000 – 30000
	grid size	40x50
	p_{cross}	0.9
	p_{mut}	0.001
Initialization parameters	$p_{operational}$	0.5
	$p(S_x = 1)$	0.5
	$p(C_x = 1)$	0.5

Table 3.1. Simulation parameters. p_{cross} is the crossover probability, p_{mut} is the mutation probability, $p_{operational}$ is the probability of a cell being operational in the initial grid, $p(S_x = 1)$ is the probability of the S_x bits of the genome equaling 1 (state-change bits, see Figure 3.12), and $p(C_x = 1)$ is the probability of the C_x bits of the genome equaling 1 (copy-rule bits, see Figure 3.12).

of evolving organisms, operating in an environment, thus engendering a coevolutionary scenario. This characteristic also holds for the cellular programming algorithm, as we shall see in Chapter 4.

We note in passing that the hardware resources required by our model only slightly exceed those of CAs. Since both models are local in nature, each cell must retain a copy of the rule in its own memory, regardless of their being identical or not. Moreover, the size of our genome is 320 bits as compared to the CA rule which requires 32 bits. Note that in this context rule copying is straightforward, requiring only a simple memory transfer. We maintain that on the scale of complexity (Section 3.1) our enhanced rule is very close to the low end, alongside the CA rule.

3.4.2   Initial results

Our first experiments were performed by running the model described above using an initial random population of rules. The parameters used are detailed in Table 3.1.

In this setup the only limitation imposed by the environment is due to the finite size of the grid, i.e., there is competition between rules for occupation of cells. The final grid obtained is one in which most cells are operational (approximately 96%). The rule population consists of different rules with some notable commonalities among them. The average value of the number of $C_C = 1$ bits in the rule genomes is approximately 31. This bit indicates whether the rule should be copied to the cell it occupies in the next time step ($C_C = 1$) or not ($C_C = 0$), and it is observed that almost all such bits in the genomes equal 1. Thus, a simple strategy has emerged which specifies that a rule, upon occupation of a certain cell, remains there, thereby preventing occupation by another rule (which can only enter a vacant cell).

Another commonality observed, among runs, was the average distribution of

C_x bits in the genomes of the rules present on the final grid. The percent of C_x bits equaling 1 is 63% and those equaling 0 is 37%. These ratios are approximately $1 - 1/e$ and $1/e$, respectively, and appeared regularly in all simulations. Since the C_x bits in the genome indicate how "active" a rule is, it is evident that activity is essential for survival, in the context of the simple scenario described. The average percentage of S_x bits equaling 1 was approximately 50%, indicating no preference for a specific state.

The results described were essentially the same for different values of the parameters in Table 3.1. One case did, however, prove slightly different, namely, $p_{mut} = 0$, i.e., using crossover alone. Here all cells in the final grid were operational with the C_C bits of all genomes equaling 1 (i.e., 32 $C_C = 1$ bits). Thus, it is evident that the initial population consists of sufficient genetic material such that perfect survivors can emerge. Mutation in this case hinders survival, however, we must bear in mind that the environment is simple and thus there appear to be no local minima which can only be avoided by using mutation. As we shall see ahead, this is not the case for more complex environments.

Another interesting phenomena was observed by looking at the $S_x = 1$ and $C_x = 1$ grids. The $S_x = 1$ grid is constructed by computing for each cell the total number of S_x bits which equal 1 for the rule genome in that cell. The $C_x = 1$ grid is constructed analogously for C_x bits. A typical run is presented in Figure 3.13, with different $S_x = 1$ and $C_x = 1$ values represented by different gray levels. It is evident that clusters are formed, according to state preference ($S_x = 1$ grid) and according to activity ($C_x = 1$ grid).

A final experiment performed in the context of the scenario described so far was the removal of the constraint that a rule may only copy itself into a vacant cell. When run with $p_{mut} = 0$, i.e., no mutations, one rule remained on the grid, occupying all cells (i.e., all cells were operational). This rule is the perfect survivor with all C_x bits in its genome set to 1.

3.4.3 Fitness in an IPD environment

In this section we enhance our model by adding a measure of a rule's *fitness*, specifying how well it performs in a certain *environment*. The environment explored is defined by the Iterated Prisoner's Dilemma (IPD), a simple game which has been investigated extensively as a model of the evolution of cooperation. IPD provides a useful framework for studying how cooperation can become established in a situation where short-range maximization of individual utility leads to a collective utility minimum. The game was first explored by Flood (1952) (see also Poundstone, 1992) and became ubiquitous due to Axelrod's work (Axelrod and Hamilton, 1981; Axelrod, 1984; Axelrod, 1987; Axelrod and Dion, 1988). These studies involve competition between several *strategies*, which are either fixed at the outset or evolve over time. An evolutionary approach was also taken by Lindgren (1992) and Lindgren and Nordahl (1994a), where genomes represent finite-memory game strategies, with an initial population containing only memory-1 strategies. The memory length is allowed to change through neutral

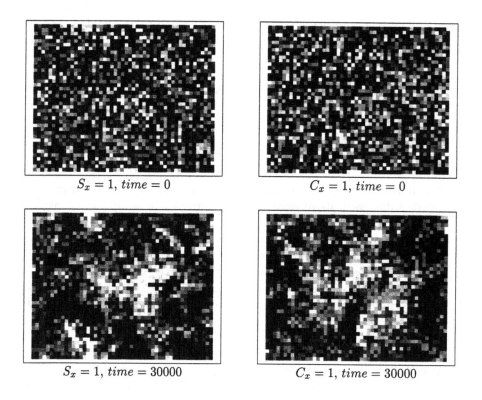

$S_x = 1,\ time = 0$ $C_x = 1,\ time = 0$

$S_x = 1,\ time = 30000$ $C_x = 1,\ time = 30000$

Figure 3.13. The $S_x = 1$ and $C_x = 1$ grids. The $S_x = 1$ grid is constructed by computing for each cell the total number of S_x bits which equal 1 for the rule genome in that cell. The $C_x = 1$ grid is constructed analogously for C_x bits. The computed values are represented by distinct gray-level values. Note that at each time step every operational cell performs its appropriate action (in accordance with its genome), after which rule evolution takes place, through application of the crossover and mutation operators.

gene duplications and split mutations, after which point mutations are applied, which can then give rise to new strategies. Simulations of this model revealed interesting phenomena of evolving strategies in a punctuated-equilibria manner (Eldredge and Gould, 1972).

The fact that the physical world has spatial dimensions has also come into play in the investigation of IPD models. A CA approach was applied by Axelrod (1984), in which each cell contains a single strategy and simultaneously plays IPD against its neighbors. The cell's score is then compared to its neighbors and the highest-scoring strategy is adopted by the cell at the next time step. In this case evolution was carried out with a fixed set of strategies, i.e., without application of genetic operators. Nowak and May (1992) considered the dynamics of two interacting memoryless strategies: cooperators and defectors (also known in the IPD literature as AllC and AllD). Spatiotemporal chaos was observed when interactions occurred on a two-dimensional grid. A spatial evolutionary model was also considered by Lindgren and Nordahl (1994b), where the representation of strategies and adaptive moves were identical to those of Lindgren (1992), described above.

It is important to note the difference of the above approaches from ours. The models discussed above were explicitly intended to study various aspects of the evolution of cooperation using the IPD game. Thus, *strategies* are the basic units of interaction, whether fixed or evolving over time (e.g., by coding them as genomes and performing genetic operators). In contrast, we use IPD to model an *environment* and our basic unit of interaction is the rule discussed in Section 3.4.1. Our genome does not represent an IPD strategy, but rather a general form of local interaction, pertinent to our model. Our intention is to study such interacting cells in various environments, one of which is defined in this section by IPD. Thus, rather than use IPD explicitly in the form of strategies, it is applied implicitly through the environment.

At each time step, every operational cell plays IPD with its neighbors, where a value of 1 represents cooperation and a value of 0 represents defection. The payoff matrix is as follows (presented for row player):

	Cooperation (1)	Defection (0)
Cooperation (1)	3	0
Defection (0)	5	1

The cell's fitness is computed as the sum of the (four) payoffs, after which the following takes place: each (operational) cell which has an operational neighbor with a higher fitness than its own "dies," i.e., becomes vacant. Crossover and mutation are then carried out as described above with one minor difference: the crossover probability p_{cross} is not fixed, but is equal to $(f(i,j) + f(i_n, j_n))/40$, where $f(i,j)$ is the fitness of the cell at position (i,j), $f(i_n, j_n)$ is the fitness of the selected operational neighbor for crossover (see Section 3.4.1). In summary, the (augmented) computational process is as follows: at each time step the grid is updated by rule application, then fitness is evaluated according to IPD, after which operational cells with fitter operational neighbors become vacant. Finally, crossover and mutation are applied as explained above.

Our simulations revealed the evolutionary phenomenon depicted in Figure 3.14 (parameters used are those of Table 3.1, except for p_{cross}, computed as discussed above). The figure presents a typical run, starting from a random grid ($time = 0$). At $time = 1050$, we observe that approximately half the cells are operational ones in state 0, surrounded by vacant cells in state 1. This configuration, which we term *alternate defection*, is one in which the operational cells attain the maximal fitness (payoff) of 20. However, this is not a stable configuration. At some point in time a small cluster of cooperating operational cells emerges ($time = 1500$), spreading rapidly throughout the grid ($time = 1650$). The final configuration is one in which most cells are cooperating operational ones with a fitness of 12 ($time = 2400$).

The notion of a cluster of cooperation in a spatial IPD model was discussed by Axelrod (1984) (albeit without rule evolution, see above). He used the term "invasion by a cluster," emphasizing that a single cooperating cell does not stand a chance against a world of defectors. As noted above, our model is more complex, involving evolutionary mechanisms and a general genome, which does not specifically code for IPD strategies. Nonetheless, we see that the IPD environment induces cooperation, with a noteworthy transition phenomenon in which widespread defection prevails.

Cooperation is achieved by a multitude of *different* rules, i.e., with different genotypic makeup. Upon inspection of these rules, we detected a significant commonality among them, found in gene g_{31}, which is usually:

$11111 \Rightarrow +++++$

or, in some cases, a C_x bit may be 0, where $x \neq c$ (i.e., not the central copy-rule bit), for example:

$11111 \Rightarrow +++1+$

Thus, we see how cooperation is maintained, by having this gene activated once stability is attained, essentially assuring that the cell remains operational and in state 1 (cooperate) with operational cooperating neighbors. Occasional "cheaters" have been observed, i.e., rules with gene g_{31} such as:

$11111 \Rightarrow -++++$

These are rules which remain operational at the next time step but in a state of defection. However, they are unsuccessful in invading the grid, and we have not observed a return to widespread defection after cooperation has been attained.

It is noteworthy that the final grid consists of rules which essentially employ only one gene of the 32 present in the genome. This may be compared to biological settings, where only part of the genome is expressed, while other parts are of no use. Thus, one of the aforementioned features of our model is demonstrated, namely, the general encoding of cellular rules, as opposed, e.g., to explicit encoding of IPD strategies. Evolution takes its course, converging to a stable "strategy," consisting of a *multitude* of *different* rules (genomes), whose commonality lies in a specific part of the genome, the part which is *expressed*, i.e.,

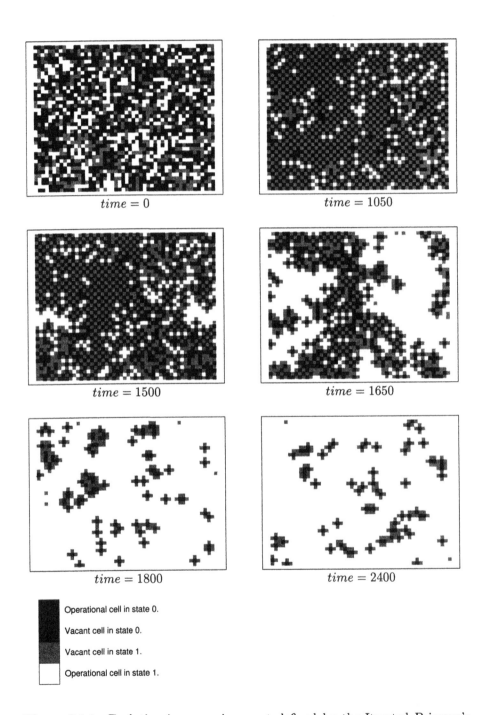

Figure 3.14. Evolution in an environment defined by the Iterated Prisoner's Dilemma.

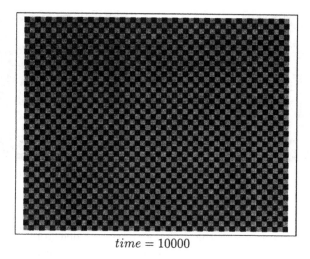

$$time = 10000$$

Figure 3.15. Evolution in the IPD environment with probability of mutation, $p_{mut} = 0$, may result in absolute alternate defection. Gray-level representation is identical to Figure 3.14.

responsible for the *phenotype*. Our rules can be viewed as simple "organisms," specified by the genome of Figure 3.12, where evolution determines which genes are expressed, along with their exact allelic form. We can view this setup as the formation of a sub-species of cooperating organisms, where members are defined by their phenotypic effects, rather than their exact genetic makeup. Whereas the genomes differ greatly (in terms of the precise alleles present), their phenotypes are similar (cooperation), due to a critical gene, g_{31}, which is the one expressed.

When p_{mut} is set to 0, two patterns have been observed to emerge: cooperation or absolute alternate defection (Figure 3.15). While cooperation is as before, among different rules, absolute alternate defection is achieved with only one surviving rule. Each such run produced a different survivor, with an important commonality found in gene g_{15}, which is one of the following:

$$01111 \Rightarrow + - - - -$$

or

$$01111 \Rightarrow 1 - - - -$$

Thus, when the grid configuration is such that all operational rules are in state 0, surrounded by vacant cells in state 1, g_{15} is activated, causing the current cell's state to become 1, and the rule to be copied into all neighboring cells, with their state changed to 0.[5] This is an interesting strategy in that an operational cell ensures cooperation of the cell it occupies and then defects to a neighboring cell.

---

[5] Note that though every vacant cell is contended by four operational neighbors they are all identical and so there is no importance as to who wins. Also note that when the center cell remains operational (as in the first g_{15} gene) it immediately dies since its fitness is 0.

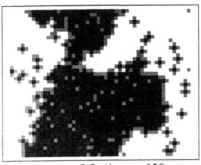
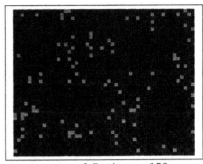

$p_{coop} = 0.9,\ time = 150$ $p_{coop} = 0.5,\ time = 150$

Figure 3.16. Evolution in the IPD environment. The initial rule population comprises only two types of rules: cooperators and defectors. The S_x bits of cooperators are set to 1, while those of defectors are set to 0. The C_x bits are initialized randomly and all cells are operational at $time = 0$. p_{coop} denotes the probability of a cell being a cooperator in the initial grid. Shown above are intermediate evolutionary phases; eventually, the grid shifts to alternate defection and then to cooperation (as in Figure 3.14).

The case of $p_{mut} = 0$ demonstrates the importance of mutation, which causes small perturbations that are necessary to invoke cooperation, as opposed to less complex environments, where mutation did not prove essential (Section 3.4.2).

We next explore the following modification: fitness is allowed to accumulate over a small period of time ($3-5$ steps). The death of operational cells still occurs at each time step as before (i.e., when a fitter operational neighbor exists), however, they stand a better chance of survival since their recent fitness histories are taken into account. It was observed that cooperation did not emerge, rather the state attained was that of alternate defection. Thus, in a "harsher" environment, inflicting immediate penalty on unfit cells, cooperation emerges, while in a more forgiving environment defection wins.

Cooperation also emerges when the grid is run with a different initial rule population, involving only two types of rules: cooperators and defectors. The S_x bits of cooperators are set to 1, while those of defectors are set to 0. The C_x bits are initialized randomly and all cells are operational at $time = 0$ (crossover and mutation are effected as above, $p_{mut} > 0$).

Let p_{coop} denote the probability of a cell being a cooperator in the initial grid. When $p_{coop} = 0.9$, we observe that at first there is a "battle" raging on between cooperators and defectors (Figure 3.16). However, the grid then shifts to alternate defection and finally to cooperation as in Figure 3.14. When p_{coop} is set to 0.5, i.e., an equal proportion (on average) of cooperators and defectors in the initial population, there is at first an outbreak of defection (Figure 3.16). Again, however, the grid shifts to alternate defection and then to cooperation.

This evolutionary pattern is also observed for $p_{coop} = 0.1$. Thus, even when there is a majority of defectors at $time = 0$, cooperation prevails.

3.4.4   Energy in an environment of niches

In this section we introduce the concept of *energy*, which serves as a measure of an organism's activity, with the intent of enhancing our understanding of phenomena occurring in our model. Each cell is considered to have a finite value of energy units. At each time step, energy units are transferred between cells in the following manner: when an operational cell attempts to copy its rule into an adjoining vacant cell an energy unit is transferred to that cell. Thus, an operational cell loses a energy units, where a equals the number of $C_x = 1$ bits, and x represents a vacant neighbor, i.e., a equals the number of copies the cell attempts to perform (not necessarily successfully since contention may occur, see Section 3.2). Note that the total amount of energy is conserved since an operational cell's loss is a vacant cell's gain. All cells hold the same amount of energy at the outset and no bounds are set on the possible energy values throughout the run.

To study the idea of energy we explore an environment consisting of spatial niches, where each cell (i, j) possesses a *niche id* equal to:

$$n_d(i, j) = \lfloor i/10 + j/10 \rfloor \bmod 5$$

The n_d value indicates the desired number of neighbors in state 1. A cell's fitness, at time t, is defined as:

$$f^t(i, j) = 4 - \mid n_d(i, j) - n_o^t(i, j) \mid$$

where $n_o^t(i, j)$ is the number of adjoining cells in state 1, at time t. As in Section 3.4.3, p_{cross} is not fixed, but is equal to $(f(i, j) + f(i_n, j_n))/8$, where $f(i_n, j_n)$ is the fitness of the selected operational neighbor. Also, an operational cell with a fitter operational neighbor "dies," i.e., becomes vacant (Section 3.4.3).

Figure 3.17 shows the grid at various times and Figure 3.18 shows the energy map, with a darker shade corresponding to lower energy. Observing the grid, it is difficult to discern the precise patterns that emerge, however, the energy map provides a clear picture of what transpires. At $time = 1000$, we note that boundaries begin to form, evident by the higher-energy borders (lighter shades). These correspond to cells positioned between niches, which remain vacant, thus becoming highly energetic. At $time = 5000$ and $time = 10000$ we see that the borders have become more pronounced. Furthermore, regions of low (dark) energy appear, corresponding to niches with $n_d = 0, 4$; this indicates that there is a lower degree of activity in these areas, presumably since these niches represent an "easier" environment. At $time = 200000$, the energy map is very smooth, indicating uniform activity, with clear borders between niches.

A different environment considered is one of temporal niches, where n_d is a function of time rather than space, with $n_d(t) = \lfloor t/1000 \rfloor \bmod 5$. We generated

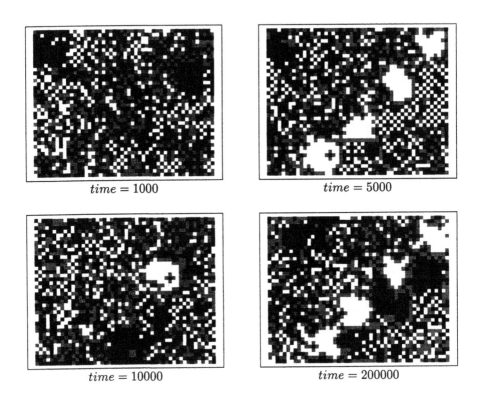

time = 1000   *time* = 5000

time = 10000   *time* = 200000

Figure 3.17. An environment defined by spatial niches: Evolution of the grid (gray-level representation is identical to Figure 3.14).

energy maps at points in time where niche shifts occur, i.e., $\lfloor t/1000 \rfloor - 1$, and observed an interesting phenomenon. After a few thousand steps the energy pattern stabilizes and the correlation between successive intervals is close to unity. Figure 3.19 depicts a typical case (for clarity we show a map of deviations from average, though the correlation was computed for the original maps). Thus, there are regions of extensive activity and regions of low activity, which persist through time.

A different aspect of the evolutionary process is considered in Figure 3.20, which shows the number of operational cells and their average fitness as a function of time. Highest fitness is obtained at temporal niches corresponding to $n_d = 4$ (*time* = 5000, 10000, 15000, 20000). At these points in time there is a drastic change in the environment (n_d shifts from 4 to 0) and we observe that fitness does not usually climb to its maximal value (which is possible for $n_d = 0$). A further observation is the correlation between fitness and operability. We see that fitness rises in exact correlation with the number of operational cells. Thus, the environment is such that more cells can become active (operational) while

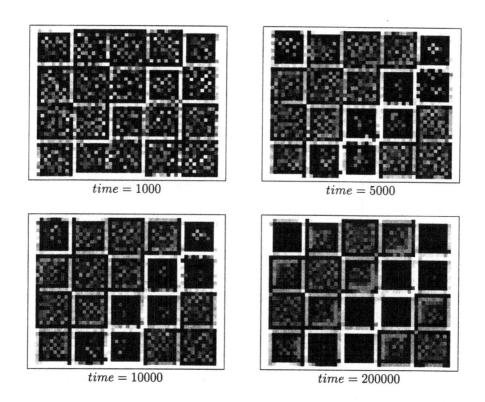

$time = 1000$ $time = 5000$

$time = 10000$ $time = 200000$

Figure 3.18. An environment defined by spatial niches: The energy map provides a clear picture of the evolutionary process, involving the formation of niches and boundaries. A darker shade corresponds to lower energy.

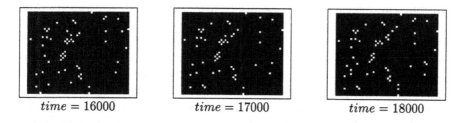

$time = 16000$ $time = 17000$ $time = 18000$

Figure 3.19. Energy evolution in an environment defined by temporal niches. Gray squares represent energy values within 2 standard deviations of the average, white squares represent extreme high values (outside the range), black squares represent extreme low values.

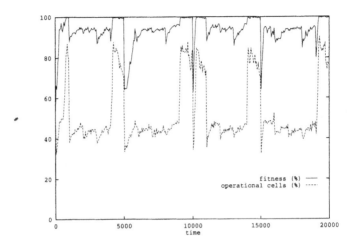

Figure 3.20. The temporal niches environment ($n_d = 0 \to 1 \to 2 \to 3 \to 4 \to$ $0 \ldots$): Fitness and operability. The number of operational cells and their average fitness (shown as percentage of maximal value), both as a function of time.

maintaining high fitness.

Such a situation is not always the case. Consider, for example, the IPD environment of Section 3.4.3, whose fitness and operability graphs are presented in Figure 3.21. Here we see that at a certain point in time fitness begins to decline, however, the number of operational cells starts rising. This is the shift from alternate defection to cooperation, discussed in Section 3.4.3. We note that in the IPD environment cells cannot all be active, while at the same time maintain the highest possible fitness. In this case lower fitness is opted for, resulting in a higher number of operational cells.

A different version of temporal niches was also studied in which n_d shifts between the values 0 and 4 every 1000 time steps. In some cases we obtained results as depicted in Figure 3.22, noting that after several thousand time steps adaptation to environmental changes becomes "easier." This could be evidence of *preadaptation*, a concept which is used to describe the process by which an organ, behavior, neural structure, etc., which evolved to solve *one* set of tasks, is later utilized to solve a *different* set of tasks. Though the concept is rooted in the work of Darwin (1866), it has more recently been elaborated by Gould (1982), Gould and Vrba (1982), and Mayr (1976).

An artificial-life approach to preadaptation was taken by Stork et al. (1992) who investigated an apparent "useless" synapse in the current tailflip circuit of the crayfish, which can be understood as being a vestige from a previous evolutionary epoch in which the circuit was used for swimming instead of flipping (as it is

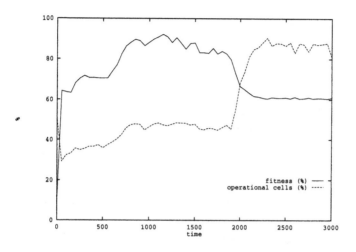

Figure 3.21. The IPD environment: Fitness and operability. The number of operational cells and their average fitness (shown as percentage of maximal value), both as a function of time.

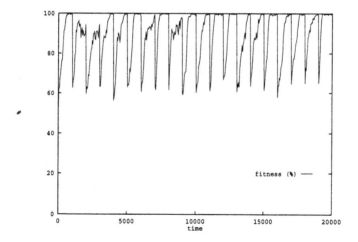

Figure 3.22. The temporal niches environment $(n_d = 0 \rightarrow 4 \rightarrow 0 \ldots)$: Average fitness as a function of time.

used today). They performed simulations in which the task of the simulated organism is switched from swimming to flipping, and then back to swimming again, observing that adaptation is much more rapid the second time swimming is selected for. This was explained in terms of evolutionary memory in which "junk" genetic information is used (Stork et al., 1992). Here "junk," stored for possible future use, is contrasted with "trash," which is discarded. Thus, apparent useless information can induce rapid fitness recovery at some future time when environmental changes occur. In the next section we examine the genescape, which allows us to directly observe the interplay of genes. One of our conclusions is that evolutionary memory can be of use since different genes are responsible for the two niches discussed above ($n_d = 0, 4$).

3.4.5   The genescape

In their paper, Bedau and Packard (1992) discuss how to discern whether or not evolution is taking place in an observed system, defining evolutionary activity as the rate at which useful genetic innovations are absorbed into the population. They point out that the rate at which new genes are introduced does not reflect genuine evolutionary activity, for the new genes may be useless. Rather, *persistent usage* of new genes is the defining characteristic of genuine evolutionary activity.

The model studied by Bedau and Packard (1992) is that of strategic bugs in which a bug's genome consists of a look-up table, with an entry for every possible combination of states. They attach to each gene (i.e., each table entry) a "usage counter," which is initialized to zero. Every time a particular table entry is used, the corresponding usage counter is incremented. Mutation sets the counter to zero, while during crossover genes are exchanged along with their counters. By keeping track of how many times each gene is invoked, waves of evolutionary activity are observed through a global histogram of gene usage plotted as a function of time. As long as activity waves continue to occur, the population is continually incorporating new genetic material, i.e., evolution is occurring (Bedau and Packard, 1992). While this measure is extremely difficult to obtain in biological settings, it is easy to do so in artificial ones, providing insight into the evolutionary process.

We have applied the idea of usage counters to our model. Each gene in our genome corresponds to a certain neighborhood configuration (input), specifying the appropriate actions to be performed (output). In this respect it is similar to the strategic bugs model of Bedau and Packard (1992) and usage counters are attached to each gene and updated as described above.[6] Bedau and Packard (1992) defined the usage distribution function, which is then used to derive the $A(t)$ measure of evolutionary activity. Since our genome is small (32 genes), we

---

[6]There is one minor difference: in the model of Bedau and Packard (1992) crossover does not occur across gene boundaries and therefore does not set the respective counter to zero, whereas in our model crossover can occur anywhere along the genome. Thus, a counter is reset whenever crossover occurs within its gene (as well as when the gene mutates).

have opted for a more direct approach in which we study the total usage of each gene throughout the grid as a function of time. For a given gene, this measure is computed by summing the usage counters of all operational cells at a given time. Our measurements can then be presented as a three-dimensional plot, denoted the *genescape*, meaning the evolutionary genetic landscape.

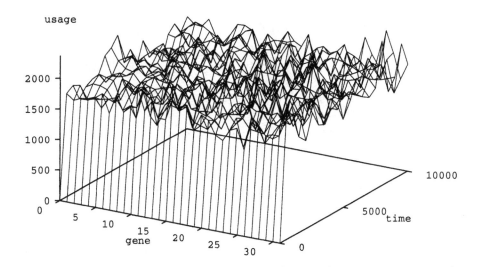

Figure 3.23. The *genescape* is a three-dimensional plot of the evolutionary genetic landscape. Essentially, it depicts the total usage throughout the entire grid of each of the 32 genes as a function of time. Shown above is the genescape for the environment of Section 3.4.2, where no explicit environmental constraints were applied.

The genescape of the environment studied in Section 3.4.2 is shown in Figure 3.23. Recall that in this case no explicit environmental constraints are placed and the only (implicit) one is therefore due to the finite size of the grid, i.e., there is competition between rules for occupation of cells. The genescape shows that usage is approximately constant (after an initial rise due to an increase in the number of operational cells) and uniform. No gene is preferred since the environment is such that all contribute equally to fitness. The constant usage value is consistent with our parameters (p_{cross} and p_{mut}). This situation may be considered as a "flat" genescape, serving as a baseline for comparison with other environments.[7]

Figure 3.24 shows the genescape of the IPD environment (Section 3.4.3). We observe that gene g_{15} initially comes to dominate, later to be overtaken by g_{31}, representing the shift from alternate defection to cooperation. Smaller peaks

---

[7]Note that other parameters did reveal interesting phenomena even for this simple environment, as noted in Section 3.4.2.

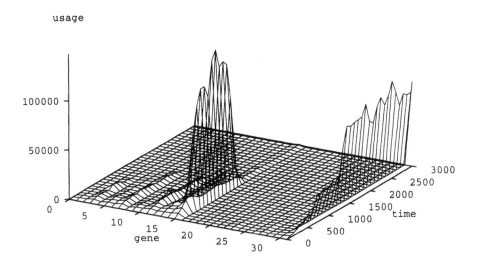

Figure 3.24. Genescape of the IPD environment of Section 3.4.3.

are also apparent, coexisting alongside g_{15}. These occur for genes g_i, such that $i < 15$, i.e., those genes representing a central cell state of defection (0). Thus, the dominance of g_{15} is not totalistic as is later the case with g_{31}. This gene, g_{31}, shows a small usage peak from the start, essentially biding its time until the "right" moment comes, when cooperation breaks through. This is reminiscent of punctuated-equilibria results, where phenotypic effects are not observed for long periods of time, while evolution runs its course in the (unobserved) genotype.

The genescapes of the temporal niches environments of Section 3.4.4 are presented in Figures 3.25 and 3.26. Observing Figure 3.25a, we note how usage peaks shift from g_0 (for niche id $n_d = 0$) to g_{31} (for $n_d = 4$) as time progresses. Closer inspection provides us with more insight into the evolutionary process (Figure 3.25b). It is noted that gene g_{16} competes with g_0 when $n_d = 0$ and g_{15} competes with g_{31} when $n_d = 4$, with g_0 and g_{31} predominating eventually. This competition is explained by the fact that n_d specifies the desired number of neighbors in state 1, without placing any restriction on the central cell, thus promoting competition between two genes, where one eventually emerges as the "winner."

When intermediate n_d values are in effect ($n_d = 1, 2, 3$), we observe multiple peaks corresponding to those genes representing the appropriate number of neighbors (Figure 3.25b). As the environment changes (through n_d), different *epistatic* effects are introduced. The lowest degree of epistasis occurs when $n_d = 0, 4$ and the highest when $n_d = 2$. It is interesting to compare these results with those obtained by Kauffman and Weinberger (1989) and Kauffman and Johnsen (1992) who employed the NK model. This model describes genotype fitness landscapes engendered by arbitrarily complex epistatic couplings. An organism's genotype

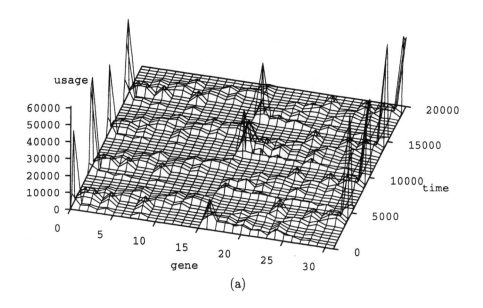

(a)

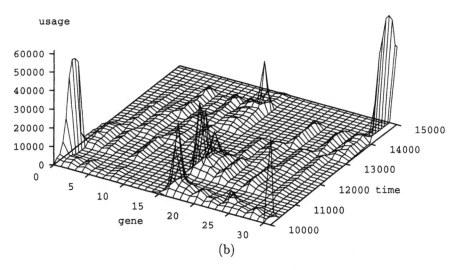

(b)

Figure 3.25. Genescape of the temporal niches environment of Section 3.4.4 $(n_d = 0 \to 1 \to 2 \to 3 \to 4 \to 0 \ldots)$. (a) Time steps $0 - 20000$. (b) Zoom of time steps $10000 - 15000$.

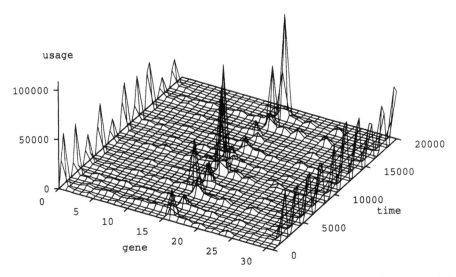

Figure 3.26. Genescape of the temporal niches environment of Section 3.4.4 $(n_d = 0 \to 4 \to 0 \ldots)$.

consists of N genes, each with A alleles. The fitness contribution of each gene depends upon itself and epistatically on K other genes. The central idea of the NK model is that the epistatic effects of the A^K different combinations of A alternative states of the other K genes on the functional contribution of the Ath state of each gene are so complex that their statistical features can be captured by assigning fitness contributions at random from a specified distribution. Tuning K from low to high increases the epistatic linkages, thus providing a tunable rugged family of model fitness landscapes.

The main conclusions offered by Kauffman and Weinberger (1989) and Kauffman and Johnsen (1992) are that as K increases relative to N (i.e., as epistatic linkages increase) the ruggedness of the fitness landscape increases by a rise in the number of fitness peaks, while the typical heights of these peaks decrease. The decrease reflects the conflicting constraints which arise when epistatic linkages increase. In the NK model epistatic linkages are made explicit via the K parameter, with fitness contributions assigned randomly. We have presented an environment in which the n_d (niche) value changes, thereby causing *implicit* changes in the degree of epistasis. Essentially, $K = 1$ for $n_d = 0, 4$, $K = 7$ for $n_d = 1, 3$, and $K = 11$ for $n_d = 2$. Our usage results of Figure 3.25 correspond to the conclusions offered by Kauffman and Weinberger (1989) and Kauffman and Johnsen (1992). As K increases, the number of usage peaks increase while their heights decrease. Note that we do not measure fitness as in the NK model, but rather usage, which can be regarded as a more "raw" measure. Also, fitness contributions are not made explicit but are rather implicitly induced by the environment. Although our viewpoint is different, the results obtained are analogous, enhancing our understanding of epistatic environmental effects.

3.4.6 Synchrony versus asynchrony

One of the prominent features of the CA model is its synchronous mode of operation, meaning that all cells are updated simultaneously at each time step. It has been observed that when asynchronous updating is used (i.e., one cell is updated at each time step), results may be different. For example, Huberman and Glance (1993) showed that when asynchrony is introduced in the model of Nowak and May (1992) (see Section 3.4.3) a fixed point is arrived at rather than the chaotic spatiotemporal behavior induced by the synchronous model. Asynchrony has also been shown to "freeze" the game of life, i.e., convergence to a fixed point occurs, rather than complex, class-IV phenomena of the synchronous model (Bersini and Detour, 1994) .

The issue raised by these investigations (see also Lumer and Nicolis, 1994) is the relevance of results obtained by CA models to biological phenomena. Indeed, Huberman and Glance (1993) have argued that patterns and regularities observed in nature require asynchronous updating since natural systems posses no global clock. It may be argued that from a physical point of view synchrony is justified: since we model a continuous spatial and temporal world, we must examine each spatial location at every time step, no matter how small we choose these (discrete) steps to be. However, as we move up the scale of complexity of the basic units, synchrony seems to be less justified. For example, IPD is usually aimed at investigating social cooperation where the basic units of interaction are complex organisms (e.g., humans, societies).

The simulations described in the previous sections were conducted using synchronous updating. Due to the arguments raised above we were motivated to investigate the issue of asynchrony by repeating some of our simulations using asynchronous updating. Results obtained were different than for synchronous updating, e.g., the asynchronous runs of the IPD environment (Section 3.4.3) produced no "interesting" configurations as for the synchronous case.

We then experimented with two forms of *partial* asynchrony: (1) *sparse updating*: at each time step a cell is updated with probability p_{sparse}, and (2) *regional updating*: at each time step a fixed-size, square region of the grid is updated. Sparse updating produced "uninteresting" results, i.e., as in the asynchronous case. However, with regional updating we observed that the synchronous updating results were repeated, provided the region size exceeded a certain value, empirically found to be approximately 100 cells (i.e., a 10x10 square).

It is noteworthy that sparse updating did not "work" even for high values of p_{sparse} (e.g., 0.2) while regional updating produced results identical to the synchronous case.[8] We also experimented with larger grids and obtained the same results without increasing the region size (10x10). While it cannot be

---

[8]Note that a region size of 10x10 is equivalent (on average) in terms of the number of cells updated per time step to $p_{sparse} = 0.05$ for a 40x50 grid.

ascertained that this size is constant, it seems safe to conjecture that it grows sub-linearly with grid size.

The regional-updating method, though not completely asynchronous, is interesting nonetheless, especially since the region size seems to grow sub-linearly with grid size. From a hardware point of view this is encouraging since implementations can maintain local (regional) synchronization, thereby facilitating scaling. We note that a minimal amount of activity must simultaneously take place in order for "interesting" patterns to emerge, i.e., there is a certain threshold of interaction. The crucial factor pertains not to the total number of cells updated per time step, but rather to the simultaneous activity of a (small) area. This is evident by the failure of the sparse-updating method as compared with the success of regional updating. The importance of "regions" of evolution has also been noted in biological settings (Mayr, 1976; Eldredge and Gould, 1972).

The issue of synchrony versus asynchrony in spatially-distributed systems is still an open question. For example, in the work of Lindgren and Nordahl (1994b), asynchronous simulations were carried out, revealing chaotic spatial organization, results which were contrasted with those of Huberman and Glance (1993). Furthermore, the work of Nowak and May (1992) was later extended by Nowak et al. (1994), showing that Huberman and Glance (1993) had only considered a specific set of parameters, and that in fact asynchronous updating does not in general induce a fixed state. Our model may yet reveal interesting phenomena for the case of complete asynchrony when other types of environments are employed. At present, we have a strong case for partial asynchrony in the form of regional updating, which, due to the small region size, is close to complete asynchrony.

3.5 Discussion

In this chapter we presented a system of simple "organisms," interacting in a two-dimensional environment, which have the capacity to evolve. We first turned our attention to designed multicellular organisms, displaying several interesting behaviors, including a self-reproducing loop, replication of passive structures by copier cells, mobile organisms, and two-phased growth and replication. These organisms offered motivation as to the power of our model in creating systems of interest. This comes about by increasing the level of operation with respect to the "physics" level of CAs.

A related work is that of *embryonics*, standing for embryonic electronics (Mange et al., 1996; Mange et al., 1995; Mange and Stauffer, 1994; Marchal et al., 1994; Durand et al., 1994). This is a CA-based approach in which three principles of natural organization are employed: multicellular organization, cellular differentiation, and cellular division. They designed an architecture which is complex enough for (quasi) universal computation, yet simple enough for physical implementation. Their approach represents another attempt at confronting the aforementioned problem of CAs, namely, the low level of operation.

An important distinction made by the embryonics group is the difference between unicellular and multicellular organisms. One of the defining characteristics of a biological cell concerns its role as the smallest part of a living being which carries the complete plan of the being, that is its genome (Mange and Stauffer, 1994). In this respect, the self-reproducing automata of von Neumann (1966) and Langton (1984) are unicellular organisms: the genome is contained within the entire configuration. An important common point between both the embryonics approach and ours is that true multicellular organisms are formed. Our cell is analogous to a biological cell in the sense that it contains the complete genome (rule table). A creature in our model consists of several cells operating in unison, thereby achieving the effect of a single "purposeful" organism. It is interesting to compare Langton's unicellular self-reproducing loop with our multicellular one (Section 3.3.1), thus illustrating our concept of raising the level of operation. Langton's loop demonstrates how unicellular replication can be attained, whereas our loop starts from there and goes on to achieve multicellular replication. In this strict sense our model may be viewed as a kind of "macro" CA, consisting of higher-level basic operations. We also observe in our model that each cell acts according to a specific gene (entry), which is a simple form of locally-based cellular differentiation. Such approaches offer new paths in the development of complex machines as collections of simpler cells. Such machines can be made to display an array of biological phenomena, including self-repair, self-reproduction, growth, and evolution (Mange and Stauffer, 1994).

After our initial investigation of multicellularity we turned our attention to evolution in rule space, which occurs through changes in the genotypes, representing the rules by which the organisms operate. At first we placed no explicit environmental constraints, thereby retaining only the implicit constraint due to the finite size of the grid. We observed that a simple strategy emerged, in which an organism (as defined by its rule) "sits tight" upon occupation of a certain cell. We can view this as the formation of simple replicators, which replicate within their own cell (at each time step), as well as into (possibly) vacant cells. It was also noted that rules tend to spatially self organize in accordance with their levels of activity (C_x bits) and state preferences (S_x bits). These results are interesting, demonstrating that even a simple environment, with but a sole constraining factor, is sufficient in order to lead the evolutionary process through regular spatiotemporal patterns.

The IPD environment revealed several interesting phenomena. The evolutionary path taken passes through a state of alternate defection, in which approximately half the cells are operational, attaining a maximal fitness. However, this is not a stable configuration, since a small cluster of cooperation eventually emerges, taking over most of the grid.

One of our observations concerns the importance of mutation in complex environments. In the simple environment of Section 3.4.2, mutation proved to be a hindrance, preventing the evolution of perfect survivors. However, as environments grew more complex, mutation became a crucial factor. For example, in the IPD environment, defection can prevail when the mutation rate is set to zero, however, cooperation always emerges when this rate is small, yet non-zero. It seems that mutation is necessary to help the evolutionary process from getting "stuck" in local minima (see also Goldberg, 1989).

The emergence of cooperation depends not only on the mutation operator but also on the "harshness" of the environment. When the environment is more forgiving, cooperation does not necessarily emerge and defection may prevail, whereas in a harsher environment defection always "steps down" in favor of co-operation. This can be compared to real-life situations, in which survival in a harsher environment may be enhanced through cooperation.

As discussed in Section 3.4.3, our IPD environment is different than other IPD models in that our genome is general and does not code for specific actions, e.g., strategies. Cooperation emerges between a multitude of different organisms, whose commonality lies in the *expression* of a specific gene, a situation which may be regarded as the formation of a sub-species.

One of the advantages of ALife models is the opportunities they offer in performing in-depth studies of the evolutionary process. This was accomplished, in our case, by observing not only phenotypic effects (i.e., cellular states as a function of time), but also by employing such measures as fitness, operability, energy, and the genescape. The energy concept was introduced as a measure of an organism's activity, where each rule copy costs one unit of energy. We applied this measure to environments consisting of spatial and temporal niches. For the case of spatial niches we observed the difficulty in discerning phenotypic effects (the grid), whereas the energy map provided us with a clear picture of the evolutionary process- regions of higher and lower activity, with high-energy boundaries between them. The environment of temporal niches presented us with an interesting phenomenon in which adaptation takes place (as evident by taking note of the fitness graph), with small clusters of extreme energetic activity forming regularly.

An additional measure introduced is the genescape, which depicts the incor-poration of new genetic material into the population. The epistatic interplay of genes is highlighted by studying such plots. In the IPD case we noted that the transition from alternate defection to cooperation occurs through a shift from one gene (g_{15}) to another (g_{31}). It was observed that while the phenotypic effect of g_{31} occurs only after several hundred time steps, it is constantly evolving, albeit at a low (dormant) rate of activity. This may provide insight on punctuated-equilibria phenomena, which could be partly explained by the difference between observed effects (phenotypes, e.g., the fossil record), and unobserved effects (genotypes).

As the environment changes through time (temporal niches), organisms adapt by traversing their adaptive landscapes. By studying the genescape we were able to observe the subtle interplay of epistatic couplings, noting shifts from single-peaked to multi-peaked, rugged terrains. Thus, we gain a deeper understanding than is possible by observing only the grid, i.e., phenotypic effects.

A tentative analogy may be put forward, between our organism and the hy-pothetical, now extinct, RNA organism (Joyce, 1989). These were presumably simple RNA molecules capable of catalyzing their own replication. What both types of organisms have in common is that a single molecule constitutes the body plus the genetic information, and effects the replication. The inherent locality and parallelism of our model add credence to such an analogy, by offering closer adherence to nature. However, we must bear in mind that only a superficial comparison may be drawn at this stage since our model is highly abstracted in relation to nature and has been implemented only for an extremely small number of "molecules." Further investigations along this line, using artificial-life models, may enhance our understanding of the RNA-world theory. The analogy between RNA organisms and other types of digital organisms has been noted by Ray (1994a).

In Section 3.1, we delineated two basic guidelines, generality and simplicity, which served us in the definition of our model. In their paper, Jefferson et al. (1992) presented a number of important properties a programming paradigm must have to be suitable as a representation for organisms in biologically-motivated studies. We discuss these below in light of our model:

1. *Computational completeness.*, i.e., Turing-machine equivalence. Since our model is an enhancement of the CA model this property holds true. We also noted that from a hardware point of view the resources required by our model only slightly exceed those of CAs (Section 3.4.1).

2. A *simple, uniform model of computation*. This is essentially what we referred to as simplicity (of basic units) and generality (the second meaning, i.e., general encoding, see Section 3.1). This property is intended to prevent the system from being biased toward a particular environment.

3. *Syntactic closure* of genetic operators. In our case all genomes represent a legal rule-table encoding. This property also enables us to start with a random population, thereby avoiding bias.

4. The paradigm should be *well conditioned* under genetic operators. This requirement is less formal, meaning that evolution between successive time steps is usually "well behaved," i.e., discontinuities occur only occasionally. This property can be assessed using the genescape.

5. *One time unit* of an organism's life must be specified. In our case time is discrete, with an organism accepting input (neighborhood states) and taking action (output) in a single time step.

6. *Scalability.* This property must be examined with some care. If we wish to add sensory apparatus (in our case, increase the neighborhood and/or the number of cellular states), then the genome grows exponentially since it encodes a finite state automaton table. However, complexity can increase through the interactions of several organisms. Indeed, a central goal of ALife research is the evolution of multicellular creatures. As noted above, such organisms are parallel devices, composed of simple basic units and

may therefore scale very well. At this point we have demonstrated that multicellularity can be attained, albeit by design (Section 3.3). Scalability is also related to the issue of asynchrony, discussed in Section 3.4.6. We shall return to this point in Section 4.7, where we describe how CAs evolved via cellular programming can be scaled.

The model presented in this chapter provides insight into issues involving adaptation and evolution. There are still, however, many limitations that should be addressed. We have modeled an environment in the strict sense (through explicit fitness definition), i.e., excluding the organisms themselves (Section 3.1). Although we achieved an environment in the broad sense, i.e., a total system of interacting organisms, the dichotomy between organisms and their environment is still a major obstacle to overcome (Jefferson et al., 1992; see also Bonabeau and Theraulaz, 1994). Another central issue discussed above is the formation (evolution) of multicellular organisms- it is clear that more research is needed in this direction.

The evolutionary studies we performed were carried out using rather small grids. It seems reasonable to assume that in order to evolve "interesting" creatures a larger number of units is required. Models such as ours, which consist of simple, locally-connected units, lend themselves to scaling through the use of parallel or distributed implementations. For example, Ray (1994b) has implemented a network-wide reserve for the digital Tierra creatures.[9] He hopes that by increasing the scale of the system by several orders of magnitude, new phenomena may arise that have not been observed in the smaller-scale systems.

It is hoped that the development of such ALife models will serve the twofold goal of: (1) increasing our understanding of biological phenomena, and (2) enhancing our understanding of artificial models, thereby providing us with the ability to improve their performance. ALife research opens new doors, providing novel opportunities to explore issues such as adaptation, evolution, and emergence, which are central both in natural environments as well as man-made ones.

In the next chapter we follow a different path, posing the question of whether parallel cellular machines can evolve to solve computational tasks. Presenting the *cellular programming* approach, we provide a positive answer to this question.

---

[9] *Tierra* is a virtual world, consisting of computer programs that can undergo evolution. In contrast to evolutionary algorithms, where fitness is defined by the user, the Tierra "creatures" (programs) receive no such direction. Rather, they compete for the "natural" resources of their computerized environment, namely, CPU time and memory. Since only a finite amount of these are available, the virtual world's natural resources are limited, as in nature, giving rise to competition between creatures. Ray observed the formation of an "ecosystem" within the Tierra world, including organisms of various sizes, parasites, and hyper-parasites (Ray, 1992).

Chapter 4

Cellular Programming: Coevolving Cellular Computation

The real voyage of discovery consists not in seeking new landscapes but in having new eyes.

Marcel Proust

Hard problems usually have immediate, brilliant, incorrect solutions.

Anonymous

4.1 Introduction

The idea of applying the biological principle of natural evolution to artificial systems, introduced more than three decades ago, has seen impressive growth in the past decade. Usually grouped under the term *evolutionary algorithms* or *evolutionary computation*, we find the domains of genetic algorithms, evolution strategies, evolutionary programming, and genetic programming (see Chapter 1). Research in these areas has traditionally centered on proving theoretical aspects, such as convergence properties, effects of different algorithmic parameters, and so on, or on making headway in new application domains, such as constraint optimization problems, image processing, neural network evolution, and more. The implementation of an evolutionary algorithm, an issue which usually remains in the background, is quite costly in many cases, since populations of candidate solutions are involved, coupled with computation-intensive fitness evaluations. One possible solution is to parallelize the process, an idea which has been explored to some extent in recent years (see reviews by Tomassini, 1996; Cantú-Paz, 1995). While posing no major problems in principle, this may require judicious modifications of existing algorithms, or the introduction of new ones, in order to meet the constraints of a given parallel machine.

In the remainder of this volume we take a different approach. Rather than ask ourselves how to better implement a specific algorithm on a given hardware platform, we pose the more general question of whether machines can be made to evolve. While this idea finds its origins in the cybernetics movement of the 1940s and the 1950s, it has recently resurged in the form of the nascent field of bio-inspired systems and evolvable hardware (Sanchez and Tomassini, 1996). The field draws on ideas from evolutionary computation as well as on recent hardware developments. The cellular machines studied are based on the simple non-uniform CA model (see Section 1.2.3 and Chapter 2), rather than the enhanced model of the previous chapter; thus, the only difference from the original CA is the non-uniformity of rules.

In this chapter we introduce the basic approach for evolving cellular machines, denoted *cellular programming*, demonstrating its viability by conducting an in-depth study of two non-trivial computational problems, density and synchronization (Sipper, 1996a; Sipper, 1997b). In the next chapter we shall describe a number of additional computational tasks, which suggest possible applications of our approach (Sipper, 1996b; Sipper, 1997a). Though most of our investigations were carried out through software simulation, one of the major goals is the attainment of "evolving ware," *evolware*, with current implementations centering on hardware, while raising the possibility of using other forms in the future, such as *bioware*. In Chapter 6, we present the "firefly" machine, an evolving, online, autonomous hardware system, based on the cellular programming approach (Goeke et al., 1997). The issue of robustness, namely, how resistant are our evolved systems in the face of errors is addressed in Chapter 7 (Sipper et al., 1996a; Sipper et al., 1996b). This can also be considered as another generalization of the original CA model (the first being non-uniformity of rules) that addresses non-deterministic CAs. Finally, in Chapter 8 we generalize on yet another aspect of CAs, namely, their standard, homogeneous connectivity, demonstrating that cellular rules can coevolve concomitantly with the cellular connections, to produce high-performance systems (Sipper and Ruppin, 1996a; Sipper and Ruppin, 1996b).

The next section presents the density and synchronization computational tasks and discusses previous work on the evolution of *uniform* CAs to solve them. Section 4.3 delineates the cellular programming algorithm used to evolve non-uniform CAs. As opposed to the standard genetic algorithm, where a population of *independent* problem solutions *globally* evolves (Section 1.3), our approach involves a grid of rules that *coevolves locally*. We next apply our algorithm to evolve non-uniform CAs to perform the density and synchronization tasks. Results using one-dimensional, radius $r = 3$ CAs are presented in Section 4.4, one-dimensional, minimal radius $r = 1$ CAs are analyzed in Section 4.5, and two-dimensional grids are employed in Section 4.6. The scalability issue is discussed in Section 4.7, where we present an algorithm for scaling evolved CAs. We end this chapter with a discussion in Section 4.8, our main conclusions being:

1. Non-uniform CAs can attain high performance on non-trivial computational tasks.

2. Such CAs can be coevolved to perform computations, with evolved, high-performance systems exhibiting quasi-uniformity.

3. Non-uniformity may reduce connectivity requirements, i.e., the use of smaller neighborhoods is made possible.

4.2 Previous work

A major impediment preventing ubiquitous computing with CAs stems from the difficulty of utilizing their complex behavior to perform useful computations. As noted by Mitchell et al. (1994b), the difficulty of designing CAs to exhibit a specific behavior or perform a particular task has severely limited their applications; automating the design (programming) process would greatly enhance the viability of CAs.

In what follows, we consider CAs that perform computations, meaning that the input to the computation is encoded as an initial configuration, the output is the configuration after a certain number of time steps, and the intermediate steps that transform the input to the output are considered to be the steps in the computation (Mitchell, 1996). The "program" emerges through "execution" of the CA rule in each cell. Note that this use of CAs as computers differs from the method of constructing a universal Turing machine in a CA, as presented in Chapter 2 (for a comparison of these two approaches see Mitchell et al., 1994a; see also Perrier et al., 1996).

The application of genetic algorithms to the evolution of *uniform* cellular automata was initially studied by Packard (1988) and recently undertaken by the EVCA (evolving CA) group (Mitchell et al., 1993; Mitchell et al., 1994a; Mitchell et al., 1994b; Das et al., 1994; Das et al., 1995; Crutchfield and Mitchell, 1995; see also review by Sipper, 1997b). They carried out experiments involving one-dimensional CAs with $k = 2$ and $r = 3$, where k denotes the number of possible states per cell and r denotes the radius of a cell, i.e., the number of neighbors on either side (thus, each cell has $2r + 1$ neighbors, including itself). Spatially periodic boundary conditions are used, resulting in a circular grid.[1] As noted in Section 1.2, a common method of examining the behavior of one-dimensional CAs is to display a two-dimensional space-time diagram, where the horizontal axis depicts the configuration at a certain time t, and the vertical axis depicts successive time steps (e.g., Figure 4.1). The term 'configuration,' formally defined in Section 1.2, refers to an assignment of 1 states to several cells, and 0s otherwise.

The EVCA group employed a standard genetic algorithm to evolve uniform CAs to perform two computational tasks, namely, density and synchronization. We first describe results pertaining to the former. The density task is to decide

---

[1]For example, for an $r = 1$ CA this means that the leftmost and rightmost cells are connected.

whether or not the initial configuration contains more than 50% 1s. Following Mitchell et al. (1993), let ρ denote the density of 1s in a grid configuration, $\rho(t)$ the density at time t, and ρ_c the threshold density for classification (in our case 0.5). The desired behavior (i.e., the result of the computation) is for the CA to relax to a fixed-point pattern of all 1s if $\rho(0) > \rho_c$, and all 0s if $\rho(0) < \rho_c$. If $\rho(0) = \rho_c$, the desired behavior is undefined (this situation shall be avoided by using odd grid sizes).

As noted by Mitchell et al. (1994b), the density task comprises a non-trivial computation for a small-radius CA ($r \ll N$, where N is the grid size). Density is a global property of a configuration, whereas a small-radius CA relies solely on local interactions. Since the 1s can be distributed throughout the grid, propagation of information must occur over large distances (i.e., $O(N)$). The minimum amount of memory required for the task is $O(\log N)$ using a serial-scan algorithm, thus the computation involved corresponds to recognition of a non-regular language. It has been proven that the density task cannot be perfectly solved by a uniform, two-state CA (Land and Belew, 1995a), however, no upper bound is currently available on the best possible imperfect performance. Note that this proof applies to the above statement of the problem, where the CA's final pattern (i.e., output) is specified as a fixed-point configuration. Interestingly, it has recently been proven that by changing the output specification, a two-state, $r = 1$ uniform CA exists, that can perfectly solve the density problem (Capcarrere et al., 1996; this result is summarized in Appendix B). Below, we shall consider the original density problem, i.e., where convergence to a fixed point is required, as it is a veritable difficult task within the CA framework, thereby posing a worthwhile "challenge" for an evolutionary algorithm.[2]

A $k = 2$, $r = 3$ rule which successfully performs this task was discussed by Packard (1988). This is the Gacs-Kurdyumov-Levin (GKL) rule, defined as follows (Gacs et al., 1978; Gonzaga de Sá and Maes, 1992):

$$s_i(t+1) = \begin{cases} \text{majority}[s_i(t), s_{i-1}(t), s_{i-3}(t)] & \text{if } s_i(t) = 0 \\ \text{majority}[s_i(t), s_{i+1}(t), s_{i+3}(t)] & \text{if } s_i(t) = 1 \end{cases}$$

where $s_i(t)$ is the state of cell i at time t.

Figure 4.1 depicts the behavior of the GKL rule on two initial configurations, $\rho(0) < \rho_c$ and $\rho(0) > \rho_c$. We observe that a transfer of information about local neighborhoods takes place to produce the final fixed-point configuration. Essentially, the rule's strategy is to successively classify local densities with the locality range increasing over time. In regions of ambiguity, a "signal" is propagated, seen either as a checkerboard pattern in space-time or as a vertical white-to-black boundary (Mitchell et al., 1993).

The standard genetic algorithm (see Section 1.3) employed by Mitchell et al. (1993) uses a randomly generated initial population of CAs, with $k = 2$, $r = 3$,

---

[2]Henceforth, unless explicitly stated, "density task" refers to the original statement of the problem, specifying a fixed-point output.

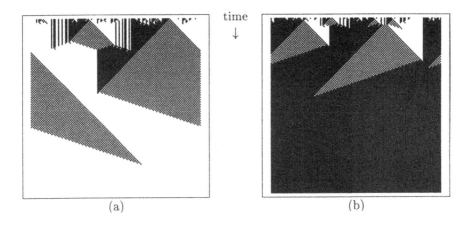

Figure 4.1.   The density task: Operation of the GKL rule.   CA is one-dimensional, uniform, 2-state, with connectivity radius $r = 3$.   Grid size is $N = 149$. White squares represent cells in state 0, black squares represent cells in state 1. The pattern of configurations is shown through time (which increases down the page). (a) Initial density of 1s is $\rho(0) \approx 0.47$ and final density at time $t = 150$ is $\rho(150) = 0$. (b) Initial density of 1s is $\rho(0) \approx 0.53$ and final density at time $t = 150$ is $\rho(150) = 1$. The CA relaxes in both cases to a fixed pattern of all 0s or all 1s, correctly classifying the initial configuration.

and a grid size of $N = 149$. Each uniform CA is represented by a bit string, delineating its rule table, containing the next-state (output) bits for all possible neighborhood configurations, listed in lexicographic order (i.e., the bit at position 0 is the state to which neighborhood configuration 0000000 is mapped to, and so on until bit 127 corresponding to neighborhood configuration 1111111). The bit string, known as the "genome," is of size $2^{2r+1} = 128$, resulting in a huge search space of size 2^{128}.

Each uniform CA (rule) in the population was run for a maximum number of M time steps, where M is selected anew for each rule from a Poisson distribution with mean 320. A rule's fitness is defined as the fraction of cell states correct at the last time step, averaged over $100 - 300$ initial configurations. At each generation a new set of configurations is generated at random, uniformly distributed over densities, i.e., $\rho(0) \in [0.0, 1.0]$. All rules are tested on this set and the population of the next generation is created by copying the top half of the current population (ranked according to fitness) unmodified; the remaining half of the next-generation population is created by applying the genetic operators of crossover and mutation to selected rules from the current population.

Using the genetic algorithm highly successful rules were found, with the best fitness values being in the range $0.93 - 0.95$. Under the above fitness function, the GKL rule has fitness ≈ 0.98. The genetic algorithm never found a rule with fitness above 0.95 (Mitchell et al., 1993; Mitchell et al., 1994b).

Another result of Mitchell et al. (1993) concerns the λ parameter, introduced by Langton (1990; 1992a) in order to study the structure of the space of CA rules. The λ of a given CA rule is the fraction of non-quiescent output states in the rule table, where the quiescent state is arbitrarily chosen as one of the possible k states. For binary-state CAs, the quiescent state is usually 0 and therefore λ equals the fraction of output-1 bits in the rule table.

In recent years it has been speculated that computational capability can be associated with phase transitions in CA rule space (Langton, 1990; Li et al., 1990; Langton, 1992a; see also Section 2.6). This phenomenon, generally referred to as the "edge of chaos," asserts that dynamical systems are partitioned into ordered regimes of operation and chaotic ones, with complex regimes arising on the edge between them. These complex regimes are hypothesized to give rise to computational capabilities. For CAs this means that there exist critical λ values at which phase transitions occur. It has been suggested that this phenomenon exists in other dynamical systems as well (Kauffman, 1993).

One of the main results of Mitchell et al. (1993) regarding the density task is that most of the rules evolved to perform it are clustered around $\lambda \approx 0.43$ or $\lambda \approx 0.57$. This is in contrast to Packard (1988), where most rules are clustered around $\lambda \approx 0.24$ or $\lambda \approx 0.83$, which correspond to λ_c values, i.e., critical values near the transition to chaos.

The results obtained by Mitchell et al. (1993) concerning the density task, coupled with a theoretical argument given in their paper, lead to the conclusion that the λ value of successful rules performing the density task is more likely to be close to 0.5, i.e., depends upon the ρ_c value. They argued that for this class of computational tasks, the λ_c values associated with an edge of chaos are not correlated with the ability of rules to perform the task. More recently, Gutowitz and Langton (1995) have reexamined this issue, suggesting that in order to find out whether there is an edge of chaos and if so, whether evolution can take us to it, one must define a good measure of complexity. They suggested that convergence time is such a measure, and demonstrated that critical rules converge on average more slowly than non-critical rules; furthermore, genetic evolution driven by convergence time produces a wide variety of complex rules. However, other results suggest that this may not always be a correct measure for a transition (Stauffer and de Arcangelis, 1996).

A study of non-uniform CA rule space has not been carried out to our knowledge. Moreover, a new parameter must be defined since a non-uniform CA contains cells with different λ values. Nonetheless, we have obtained results which lend support to the conclusions of Mitchell et al. (1993) (see Section 4.4).

The second task investigated by Das et al. (1995) is the synchronization task: given any initial configuration, the CA must reach a final configuration, within M time steps, that oscillates between all 0s and all 1s on successive time steps.

As noted by Das et al. (1995), this is perhaps the simplest, non-trivial synchronization task.

The task is non-trivial since synchronous oscillation is a global property of a configuration, whereas a small-radius CA employs only local interactions. Thus, while local regions of synchrony can be directly attained, it is more difficult to design CAs in which spatially-distant regions are in phase. Since out-of-phase regions can be distributed throughout the lattice, transfer of information must occur over large distances (i.e., $O(N)$) to remove these phase defects and produce a globally synchronous configuration. Das et al. (1995) reported that in 20% of the runs the genetic algorithm discovered successful CAs with a maximal fitness value of 1.

The phenomenon of synchronous oscillations occurs in nature, a striking example of which is exhibited by fireflies. Thousands such creatures may flash on and off in unison, having started from totally uncoordinated flickerings (Buck, 1988). Each insect has its own rhythm, which changes only through local interactions with its neighbors' lights. Another interesting case involves pendulum clocks: when several of these are placed near each other, they soon become synchronized by tiny coupling forces transmitted through the air or by vibrations in the wall to which they are attached (for a review on other phenomena in nature involving synchronous oscillation see Strogatz and Stewart, 1993).

In the next section we present the cellular programming algorithm, and apply it in the remainder of this chapter to the evolution of non-uniform CAs to perform the aforementioned tasks.

4.3 The cellular programming algorithm

We study 2-state, non-uniform CAs, in which each cell may contain a different rule. A cell's rule table is encoded as a bit string (the "genome"), containing the next-state (output) bits for all possible neighborhood configurations (as explained in Section 4.2). Rather than employ a *population* of evolving, uniform CAs, as with the standard genetic algorithm, our algorithm involves a *single*, non-uniform CA of size N, with cell rules initialized at random.[3] Initial configurations are then generated at random, in accordance with the task at hand, and for each one the CA is run for M time steps (in our simulations we used $M \approx N$ so that computation time is linear with grid size). Each cell's *fitness* is accumulated over $C = 300$ initial configurations, where a single run's score is 1 if the cell is in the correct state after M time steps, and 0 otherwise. After every C configurations evolution of rules occurs by applying crossover and mutation. This evolutionary process is performed in a completely *local* manner, where genetic operators are

---

[3]To increase rule diversity in the initial grid, the rule tables were randomly selected so as to be uniformly distributed among different λ values. Note that our algorithm is not necessarily restricted to a single, non-uniform CA since an ensemble of distinct grids can evolve independently in parallel.

applied only between directly connected cells. It is driven by $nf_i(c)$, the number of fitter neighbors of cell i after c configurations. The pseudo-code of our algorithm is delineated in Figure 4.2. In our simulations, the total number of initial configurations per evolutionary run was in the range $[10^5, 10^6]$.[4]

Crossover between two rules is performed by selecting at random (with uniform probability) a single crossover point, and creating a new rule by combining the first rule's bit string before the crossover point with the second rule's bit string from this point onward. Mutation is applied to the bit string of a rule with probability 0.001 per bit.

There are two main differences between our algorithm and the standard genetic algorithm (Section 1.3): (1) The latter involves a population of evolving, uniform CAs, with all individuals *ranked* according to fitness, and crossover occurring between *any* two individuals in the population. Thus, while the CA runs in accordance with a local rule, evolution proceeds in a *global* manner. In contrast, our algorithm proceeds *locally* in the sense that each cell has access only to its locale, not only during the run but also during the evolutionary phase, and no global fitness ranking is performed. As we shall see in Chapter 6, this characteristic is of prime import where hardware implementation is concerned. (2) The standard genetic algorithm involves a population of *independent* problem solutions, meaning that the CAs in the population are assigned fitness values independent of one another, and interact only through the genetic operators in order to produce the next generation. In contrast, our CA *coevolves* since each cell's fitness depends upon its evolving neighbors (the coevolutionary aspect was also noted for our ALife model in Section 3.4.1).[5]

This latter point comprises a prime difference between our algorithm and parallel genetic algorithms, which have attracted attention over the past few years. These aim to exploit the inherent parallelism of evolutionary algorithms, thereby decreasing computation time and enhancing performance (Tomassini, 1996; Cantú-Paz, 1995; Tomassini, 1995). A number of models have been suggested, among them coarse-grained, island models (Starkweather et al., 1991; Cohoon et al., 1987; Tanese, 1987), and fine-grained, grid models (Tomassini, 1993; Manderick and Spiessens, 1989). The latter resemble our system in that they are massively parallel and local, however, the coevolutionary aspect is missing. As we wish to attain a system displaying global computation, the individual cells do not evolve independently, as with genetic algorithms (be they parallel or serial), i.e., in a "loosely-coupled" manner, but rather coevolve, thereby comprising a "tightly-coupled" system.

Note that in the case of uniform CAs a *single* rule is sought, which must be universally applied to all cells in the grid, a task which may be arduous even for

---

[4]By comparison, Mitchell et al. (1993; 1994b) employed a genetic algorithm with a population size of 100, which was run for 100 generations. Every generation, each CA was run on $100 - 300$ initial configurations, resulting in a total of $[10^6, 3 \cdot 10^6]$ configurations per evolutionary run.

[5]This may also be considered a form of symbiotic cooperation, which falls, as does coevolution, under the general heading of "ecological" interactions (see Mitchell, 1996, pages 182-183).

for each cell i in CA **do in parallel**
   initialize rule table of cell i
   $f_i = 0$ { fitness value }
end parallel for
$c = 0$ { initial configurations counter }
while not done **do**
   generate a random initial configuration
   run CA on initial configuration for M time steps
   for each cell i **do in parallel**
      if cell i is in the correct final state **then**
         $f_i = f_i + 1$
      end if
   end parallel for
   $c = c + 1$
   if c mod $C = 0$ **then** { evolve every C configurations}
      for each cell i **do in parallel**
         compute $nf_i(c)$ { number of fitter neighbors }
         if $nf_i(c) = 0$ **then** rule i is left unchanged
         else if $nf_i(c) = 1$ **then** replace rule i with the fitter neighboring rule,
                 followed by mutation
         else if $nf_i(c) = 2$ **then** replace rule i with the crossover of the two fitter
                 neighboring rules, followed by mutation
         else if $nf_i(c) > 2$ **then** replace rule i with the crossover of two randomly
                 chosen fitter neighboring rules, followed by mutation
                 (this case can occur if the cellular neighborhood includes
                 more than two cells)
         end if
         $f_i = 0$
      end parallel for
   end if
end while

Figure 4.2. Pseudo-code of the cellular programming algorithm.

an evolutionary algorithm. For non-uniform CAs, search-space sizes are vastly larger than for uniform ones, a fact which initially seems as an impediment. Nonetheless, as we shall see below, good results are obtained, i.e., successful systems can be coevolved.

4.4   Results using one-dimensional, $r = 3$ grids

In this section we apply the cellular programming algorithm to evolve non-uniform, radius $r = 3$ CAs to perform the density task;[6] this cellular neighborhood is identical to that studied by Packard (1988) and Mitchell et al. (1993). For our algorithm we used randomly generated initial configurations, uniformly distributed over densities in the range $[0.0, 1.0]$, with the size $N = 149$ CA being run for $M = 150$ time steps (as noted above, computation time is thus linear with grid size). Fitness scores are assigned to each cell, ensuing the presentation of each initial configuration, according to whether it is in the correct state after M time steps or not (as described in the previous section).

One of the measures we shall report upon below is that of *performance*, defined as the average fitness of all grid cells over the last C configurations, normalized to the range $[0.0, 1.0]$. Before proceeding, we point out that this is somewhat different than the work of Mitchell et al. (1993; 1994b) on the density task, where three measures were defined: (1) performance- the number of correct classifications on a sample of initial configurations, randomly chosen from a binomial distribution over initial densities, (2) performance fitness- the number of correct classifications on a sample of C initial configurations, chosen from a uniform distribution over densities in the range $[0.0, 1.0]$ (no partial credit is given for partially-correct final configurations), and (3) proportional fitness- the fraction of cell states correct at the last time step, averaged over C initial configurations, uniformly distributed over densities in the range $[0.0, 1.0]$ (partial credit is given). Our performance measure is analogous to the latter measure, however, there is an important difference: as our evolutionary algorithm is local, fitness values are computed for each individual cell; global fitness of the CA can then be *observed* by averaging these values over the entire grid. As for the choice of initial configurations, Mitchell et al. (1993; 1994b) remarked that the binomial distribution is more difficult than the uniform-over-densities one since the former results in configurations with a density in the proximity of 0.5, thereby entailing harder correct classification. This distinction did not prove significant in our studies since it essentially concerns only the density task, and does not pertain to the other tasks studied in this volume. We therefore selected the uniform-over-densities distribution as a benchmark measure by which to evolve CAs and compare their performance. We shall, nonetheless, demonstrate in Chapter 8 coevolved CAs that attain high

---

[6]For the synchronization task, perfect results were obtained, both for $r = 3$ CAs, as well as for minimal radius, $r = 1$ ones, discussed in the next section. We therefore concentrate on such minimal grids for this task.

performance on the density task, even when applying the binomial distribution.

Figure 4.3 displays the results of two typical runs. Each graph depicts the average fitness of all grid cells after C configurations (i.e., performance), and the best cell's fitness, both as a function of the number of configurations run so far. Figure 4.3a displays a successful run in which the performance reaches a peak value of 0.92. Some runs were unsuccessful, i.e., performance did not rise above 0.5, the expected random value (Figure 4.3b). Observing Figure 4.3a, we note how a successful run comes about: at an early stage of evolution, a high-fitness cell rule is "discovered," as evident by the top curve depicting best fitness. Such a rule is sufficient in order to "drive" evolution in a direction of improved performance. The threshold fitness that must be attained by this rule is approximately 0.65; a lower value is insufficient to drive evolution "upwards."

We next performed an experiment in which the CA with the highest performance in the run is saved[7] and then tested on 10,000 initial configurations, generated as detailed above. A slight drop in fitness was observed, though the value is still high at approximately 0.91.

The success of our cellular programming algorithm in finding good solutions, i.e., high-performance, non-uniform CAs, is notable if one considers the search space involved. Since each cell contains one of 2^{128} possible rules and there are $N = 149$ such cells, our space is of size $(2^{128})^{149} = 2^{19072}$. This is vastly larger than uniform CA search-space sizes, nonetheless, using a local, coevolutionary algorithm, non-uniform CAs are discovered, which exhibit high performance.

A histogram of the total number of evolved rules as a function of λ is presented in Figure 4.4. It is clear that rules are clustered in the vicinity of $\lambda = 0.5$, mostly in the region $0.45 - 0.55$. As noted in Section 4.2, a study of non-uniform CA rule space has not been carried out to our knowledge. Nonetheless, we believe that this result lends support to the conclusion of Mitchell et al. (1993), stating that λ values for the density task are likely to be close to 0.5.

4.5 Results using one-dimensional, $r = 1$ grids

The work of Packard (1988), Mitchell et al. (1993; 1994b), and Das et al. (1995), concentrated solely on uniform CAs with $r = 3$, i.e., seven neighbors per cell. In this section we examine minimal radius, $r = 1$, non-uniform CAs, asking if they can attain high performance on the density and synchronization tasks, and whether such CAs can be coevolved using cellular programming. With $r = 1$, each cell has access only to its own state and that of its two adjoining neighbors. Note that *uniform*, one-dimensional, 2-state, $r = 1$ CAs are not computation universal (Lindgren and Nordahl, 1990) and do not exhibit class-IV behavior (Wolfram, 1983; Wolfram, 1984b).

---

[7]This entails saving the ensemble of rules, discovered during the evolutionary run, comprising the non-uniform CA with the highest performance (as in Appendix C).

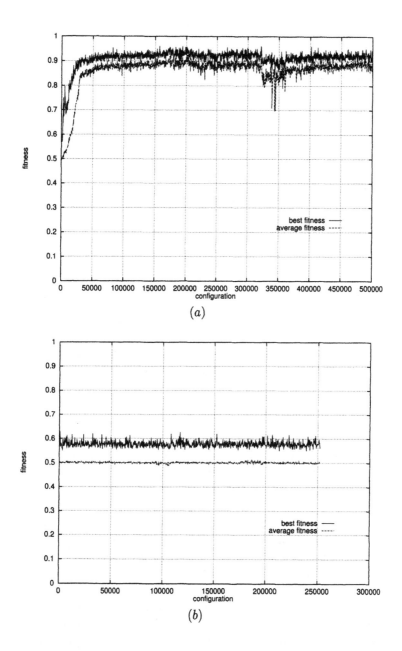

Figure 4.3. The density task ($r = 3$): Results of two typical evolutionary runs.
(a) A successful run. (b) An unsuccessful run. In both graphs, the bottom curve
depicts performance, i.e., the average fitness of all grid cells (rules), and the top
curve depicts the fitness of the best cell (rule) in the grid, both as a function of
the number of configurations presented so far.

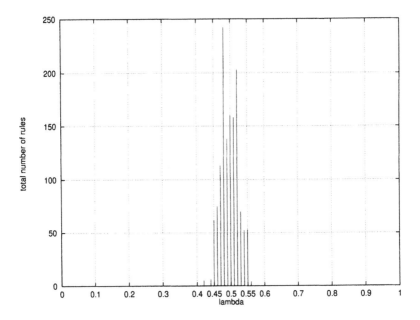

Figure 4.4. The density task ($r = 3$): Histogram depicting the total number of evolved rules (summed over several runs) as a function of λ.

4.5.1   The density task

We first describe results pertaining to the density task. The search space for non-uniform, $r = 1$ CAs is extremely large: since each cell contains one of 2^8 possible rules, and there are $N = 149$ such cells, our space is of size $(2^8)^{149} = 2^{1192}$. In contrast, the size of *uniform*, $r = 1$ CA rule space is small, consisting of only $2^8 = 256$ rules. This enables us to test each and every one of these rules on the density task, a feat not possible for larger values of r. We performed several runs in which each uniform rule was tested on 1000 random initial configurations. Results are depicted in Figure 4.5 as fitness versus rule number.[8] The fitness (performance) measure is identical to that used above for non-uniform CAs, i.e., the number of cells in the correct state after M time steps, averaged over the presented configurations (as before, fitness is normalized to the range $[0.0, 1.0]$). Rule numbers are given in accordance with Wolfram's convention (Wolfram, 1983), representing the decimal equivalent of the binary number encoding the rule table (Figure 4.6).

Having "charted" the performance of all uniform, $r = 1$ CAs, we observe that the highest value is 0.83. This is attained by rule 232 which performs a majority vote among the three neighborhood states. Thus, the maximal performance

---

[8]Results of one run are displayed- no significant differences were detected among runs.

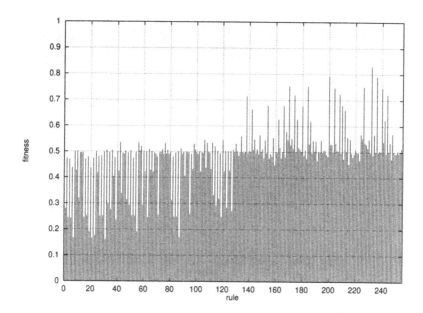

Figure 4.5. The density task: Fitness results of all possible uniform, $r = 1$ CA rules.

				Rule 224										Rule 226				
	7	6	5	4	3	2	1	0		7	6	5	4	3	2	1	0
neighborhood	111	110	101	100	011	010	001	000		111	110	101	100	011	010	001	000
output bit	1	1	1	0	0	0	0	0		1	1	1	0	0	0	1	0

				Rule 232										Rule 234				
	7	6	5	4	3	2	1	0		7	6	5	4	3	2	1	0
neighborhood	111	110	101	100	011	010	001	000		111	110	101	100	011	010	001	000
output bit	1	1	1	0	1	0	0	0		1	1	1	0	1	0	1	0

Figure 4.6. Rules involved in the density task ($r = 1$).

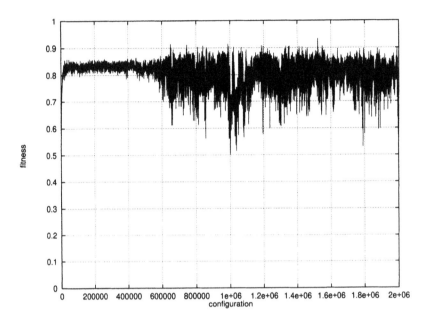

Figure 4.7. The density task: result of a typical run with $r = 1$. The graph depicts performance, i.e., the average fitness of all grid cells, as a function of the number of configurations presented so far.

of uniform, $r = 1$ CAs on the density task is known, and we now ask whether non-uniform CAs can be coevolved to attain higher performance. Observing Figure 4.7, depicting the progression of a typical run, we note that peak performance reaches a value of 0.93. Thus, we find that indeed high-performance, non-uniform CAs can coevolve, surpassing all uniform, $r = 1$ ones.

How do our evolved, non-uniform CAs manage to outperform the best uniform CA? Figure 4.8 demonstrates the operation of two uniform CAs, rule 232 (majority), which is the best uniform CA rule with fitness 0.83, and rule 226, which has fitness 0.75 (rules are delineated in Figure 4.6). We note that rule 232 exhibits a small amount of information transfer during the first few time steps, however, it quickly settles into an (incorrect) fixed point. Rule 226 shows patterns of information transfer similar to those observed with the GKL rule (Figure 4.1), however, no "decision" is reached.

The evolved, non-uniform CAs consist of a grid in which one rule dominates, i.e., occupies most grid cells; these are quasi-uniform, type-2 grids, as defined in Section 2.5.[9] In the lower-performance CAs the dominant rule is 232 (majority), whereas in the high-performance CAs rule 226 has gained dominance. We noted

---

[9] The definitions in Section 2.5 involve the limit of the given ratios, as grid size tends to infinity. For finite grids, quasi-uniformity simply implies that these ratios are small.

Figure 4.8. The density task: Operation of uniform, $r = 1$ rules. Grid size is $N = 149$. (a) Rule 232 (majority). $\rho(0) \approx 0.40$, $\rho(150) \approx 0.32$. (b) Rule 232. $\rho(0) \approx 0.60$, $\rho(150) \approx 0.66$. (c) Rule 226. $\rho(0) \approx 0.40$, $\rho(150) \approx 0.40$. (d) Rule 226. $\rho(0) \approx 0.60$, $\rho(150) \approx 0.60$.

above that rule 226 attains a fitness of only 0.75 on its own, though it is better than rule 232 at information transfer. Evolution led our non-uniform CAs toward a grid in which *most but not all* of the cells contain rule 226. Thus, instead of a low-performance uniform CA, evolution has found a high-performance *quasi-uniform* CA, with one dominant rule occupying most grid cells, while the other cells contain other rules.

The operation of one such typical CA is demonstrated in Figure 4.9, along with a rules map, depicting the distribution of rules by assigning a unique gray level to each distinct rule. The grid consists of 146 cells containing rule 226, 2 cells containing rule 224, and 1 cell containing rule 234.[10] Rules 224 and 234 differ by one bit from rule 226 (Figure 4.6). In Figure 4.9a, $\rho(0) \approx 0.40$, and we note that behavior is different at the two cells located about one third from the right, which contain rule 224. This rule maps neighborhood 001 to state 0 instead of state 1, as does rule 226, thus enhancing a neighborhood with a majority of 0s. The cells act as "buffers," which prevent erroneous information from flowing across them. In Figure 4.9b, $\rho(0) \approx 0.60$, and the cell located near the right side of the grid, containing rule 234, acts as a buffer. This rule maps neighborhood 011 to 1 instead of 0, as does rule 226, thus enhancing a neighborhood with a majority of 1s.

The uniform, rule-226 CA is capable of global information transfer, however, erroneous decisions are reached. The non-uniform CA uses the capability of rule 226, by inserting buffers in order to prevent information from flowing *too* freely. The buffers make local corrections to the signals, which are then enhanced in time, ultimately resulting in a correct output. Thus, an evolved, quasi-uniform CA outperforms a uniform one.

Further insight may be gained by studying the *genescape* of our problem, i.e., the three-dimensional plot depicting the evolutionary genetic landscape (Section 3.4.5). The cellular genome for $r = 1$ CAs is small, consisting of only 8 "genes," i.e., rule table entries.

The genescape of the non-uniform CA of Figure 4.9 is presented in Figure 4.10. We observe that genes 0 and 7 are the ones used most extensively. These correspond to neighborhoods 000 (which is mapped to 0 by all grid rules) and 111 (which is mapped to 1). This demonstrates the preservation of correct local information. Genes 4 (neighborhood 100) and 6 (neighborhood 110), of intermediate usage, also act to preserve the current state.

Observing Figure 4.10, we note that after several thousand configurations in which the landscape between genes 0 and 7 is mostly flat, a growing ridge appears, reflecting the increasing use of genes 2 and 5. These correspond to neighborhoods 010 (which is mapped to 0) and 101 (which is mapped to 1). The use of these genes changes the state of the central cell, reflecting an incorrect

---

[10]In Chapter 7 we shall take a closer look at the effects of rules distribution within the grid. Appendix C completely specifies a number of non-uniform CAs, coevolved to solve the density and synchronization tasks. The one dubbed 'Density 2' is the CA discussed in this paragraph.

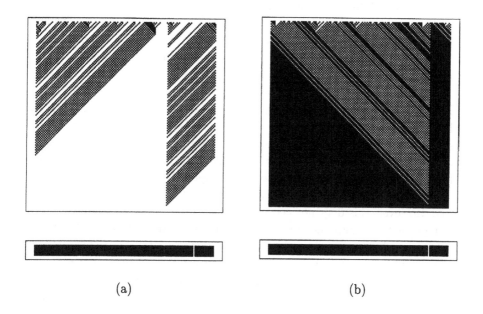

(a)   (b)

Figure 4.9. The density task: Operation of a coevolved, non-uniform, $r = 1$ CA. Grid size is $N = 149$. Top figures depict space-time diagrams, bottom figures depict rule maps. (a) $\rho(0) \approx 0.40$, $\rho(150) = 0$. (b) $\rho(0) \approx 0.60$, $\rho(150) = 1$.

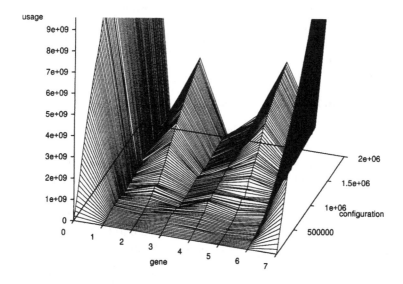

Figure 4.10. The density task: Genescape of a coevolved, non-uniform, $r = 1$ CA.

state with respect to the local neighborhood. Essentially, these two genes are related to information transfer, which increases as evolution proceeds. The least used genes are 1 and 3, which are exactly those genes by which the dominant rule (226) differs from the other two (224, 234). Though used sparingly, they are crucial to the success of the system, as noted above.

Returning to Figure 4.7, we observe that the evolutionary pattern consists of a period of relative stability followed by a period of instability, with the high-performance CA found during the unstable period. This pattern resembles the punctuated-equilibria phenomenon observed in nature (Eldredge and Gould, 1972) and also in artificial-life experiments (Lindgren, 1992; Lindgren and Nordahl, 1994a; Sipper, 1995c; see also Chapter 3). It is noted that for $r = 1$ the rule table contains only 8 bits and therefore 1-bit changes may have drastic effects. However, this does not account for the initial period of stability. Moreover, note that the graph depicts average fitness and therefore instability is due to a system-wide phenomenon taking place. Fitness variations of a small number of cells would not be sufficient to create the observed evolutionary pattern.

The configuration at which the usage ridge begins its ascent (Figure 4.10) coincides with the beginning of the unstable period (Figure 4.7). As argued above, the ridge represents the growing use of information transfer. Thus, the shift from stability to instability may represent a phase transition for our evolving CAs. As with the edge of chaos, such a shift may be associated with computational capability, evident in our case by increased computational performance. While a full explanation of the punctuated-equilibria phenomenon does not yet exist, evidence has been mounting as to its ubiquity among various evolutionary systems. Our above results suggest a possible connection between this phenomenon and the edge of chaos (see also Kauffman, 1993, page 269, on the relation between coevolution and punctuated equilibria).

4.5.2 The synchronization task

In this section we apply the cellular programming algorithm to the evolution of non-uniform, $r = 1$ CAs to perform the synchronization task. As noted in Section 4.5.1, the size of *uniform*, $r = 1$ CA rule space is small, consisting of only $2^8 = 256$ rules. This enables us to test each and every one of these rules on the synchronization task, results of which are depicted in Figure 4.11. The fitness (performance) value is computed as follows: if the density of 1s after M time steps is greater than 0.5 (respectively, less than 0.5) then over the next four time steps (i.e., $[M + 1..M + 4]$) each cell should exhibit the sequence of states $0 \rightarrow 1 \rightarrow 0 \rightarrow 1$ (respectively, $1 \rightarrow 0 \rightarrow 1 \rightarrow 0$). This measure builds upon the configuration attained by the CA at the last time step (as with the density task, fitness is averaged over several initial configurations and the final value is normalized to the range $[0.0, 1.0]$). Note that this fitness value is somewhat different than that used in Section 6.3 for the firefly machine.

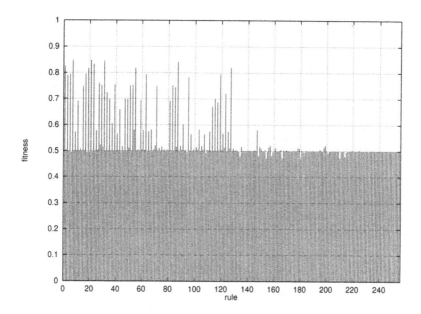

Figure 4.11. The synchronization task: Fitness results of all possible uniform, $r = 1$ CA rules.

The highest uniform performance value is 0.84, and is attained by a number of rules. These map neighborhood 111 to 0 and neighborhood 000 to 1; they also map either 110 to 0 or 001 to 1 (or both). Figure 4.12 depicts the operation of two such uniform rules. Thus, the maximal performance of uniform, $r = 1$ CAs on the synchronization task is known, and we now ask whether non-uniform CAs can attain higher performance, and whether such CAs can be found using the cellular programming algorithm.

Our algorithm yielded (on most runs) non-uniform CAs with a fitness value of 1, i.e., perfect performance is attained.[11] Again, we note that non-uniform CAs can indeed be coevolved to attain high performance, surpassing that of uniform, $r = 1$ CAs. Figure 4.13 depicts the operation of two such coevolved CAs.

The coevolved, non-uniform CAs are quasi-uniform, type-1 (Section 2.5). The number of rules in the final grids is small, ranging (among different evolutionary runs) between 3–9, with 2–3 dominant rules. Each run produced a different ensemble of rules (comprising the non-uniform CA), all of which are higher performance ones in the uniform case (Figure 4.11).[12]

---

[11]The term 'perfect' is used here in a stochastic sense since we can neither exhaustively test all 2^{149} possible configurations, nor are we in possession to date of a formal proof. Nonetheless, we have tested our best-performance CAs on numerous configurations, for all of which synchronization was attained.

[12]For example, the non-uniform CA whose operation is depicted in Figure 4.13a consists of

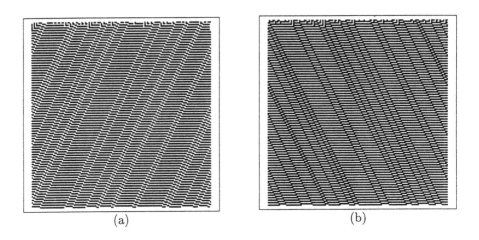

(a)   (b)

Figure 4.12. The synchronization task: Operation of uniform, $r = 1$ rules. Grid size is $N = 149$. Initial configurations were generated at random. (a) Rule 21. (b) Rule 31.

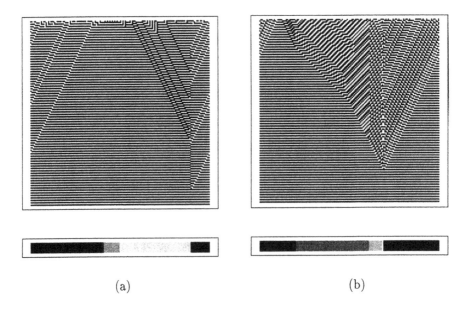

(a)   (b)

Figure 4.13. The synchronization task: Operation of two coevolved, non-uniform, $r = 1$ CAs. Grid size is $N = 149$. Top figures depict space-time diagrams, bottom figures depict rule maps.

As opposed to the uniform CAs of Figure 4.12, where no useful information transfer is evident, we note in Figure 4.13 that signals appear, as in the GKL rule (Figure 4.1). The organization of rules in the coevolved CAs is not random, but rather consists of consecutive blocks of cells containing the same rule. This can be noted by observing the rule maps depicted in Figure 4.13. Comparing these maps with the corresponding space-time diagrams, we note that signals interact at the boundaries between rule blocks, thereby ultimately achieving synchronization. This resembles the particle-based computation, discussed by Das et al. (1994; 1995) and by Crutchfield and Mitchell (1995). They analyzed CA computation in terms of "particles," which are a primary mechanism for carrying information ("signals") over long space-time distances. This information might indicate, for example, the result of some local processing that has occurred at an earlier time.

The uniform CAs investigated by the EVCA group have radius $r = 3$. We have found that non-uniform, $r = 1$ CAs can attain high performance, surpassing uniform CAs of the same radius. Uniform, two-dimensional, 2-state, 5-neighbor CAs cannot attain universal computation (Chapter 2), as also uniform, one-dimensional, 2-state, $r = 1$ CAs (Lindgren and Nordahl, 1990). Quasi-uniform CAs are capable of universal computation (Chapter 2; see also Sipper, 1995b), and we have demonstrated above that evolution on non-trivial tasks has found quasi-uniform CAs. These results suggest that non-uniformity reduces connectivity requirements, i.e., the use of smaller radiuses is made possible. More generally, we maintain that non-uniform CAs offer new paths toward complex computation.

4.6   Results using two-dimensional, 5-neighbor grids

Both the density and synchronization tasks can be extended in a straightforward manner to two-dimensional grids.[13] Applying our algorithm to the evolution of such CAs to perform the density task yielded notably higher performance than the one-dimensional case, with peak values of 0.99. Moreover, computation time, i.e., the number of time steps taken by the CA until convergence to the correct final pattern, is shorter (we shall elaborate upon these issues in Chapter 8). Figure 4.14 demonstrates the operation of one such coevolved CA. Qualitatively, we observe the CA's "strategy" of successively classifying local densities, with the locality range increasing over time. "Competing" regions of density 0 and density 1 are manifest, ultimately relaxing to the correct fixed point. For the synchronization task, perfect performance was evolved for two-dimensional CAs (as for the one-dimensional case). One such CA is depicted in Figure 4.15.

---

the following rules (Appendix C): rule 21 (appears 74 times in the grid), rule 31 (59 times), rule 63 (13 times), rule 53 (2 times), and rule 85 (1 time). All these rules have uniform fitness values above 0.7 (Figure 4.11).

[13]Spatially periodic boundary conditions are applied, resulting in a toroidal grid.

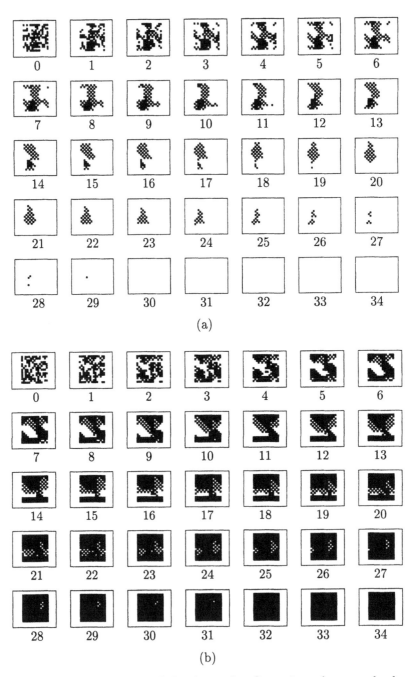

Figure 4.14. Two-dimensional density task: Operation of a coevolved, non-uniform, 2-state, 5-neighbor CA. Grid size is $N = 225$ (15×15). Numbers at bottom of images denote time steps. (a) Initial density of 1s is 0.49, final density is 0. (b) Initial density of 1s is 0.51, final density is 1.

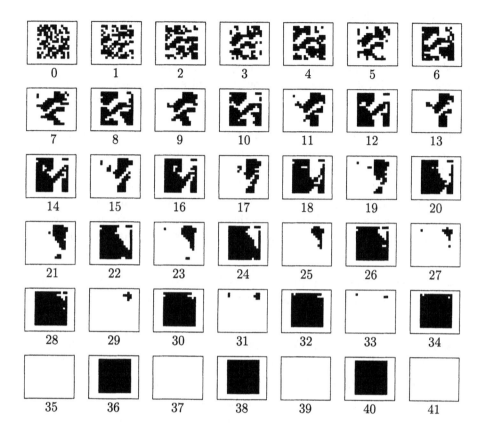

Figure 4.15. Two-dimensional synchronization task: Operation of a coevolved, non-uniform, 2-state, 5-neighbor CA. Grid size is $N = 225$ (15×15). Numbers at bottom of images denote time steps.

4.7   Scaling evolved CAs

In this section we consider one aspect of the scalability issue, which essentially involves two separate matters: the evolutionary algorithm and the evolved solutions. As to the former, we note that as our cellular programming algorithm is local, it scales better in terms of hardware resources than the standard (global) genetic algorithm. Adding grid cells requires only local connections in our case, whereas the standard genetic algorithm includes global operators such as fitness ranking and crossover (this point shall prove to be of prime import for the hardware implementation discussed in Chapter 6). In this section we concentrate on the second issue, namely, how can the grid size be modified given an evolved grid of a particular length, i.e., how can evolved solutions be scaled? This has been purported as an advantage of uniform CAs, since one can directly use the evolved rule in a grid of any desired size. However, this form of *simple* scaling does not bring about *task* scaling; as demonstrated, e.g., by Crutchfield and Mitchell (1995) for the density task, performance decreases as grid size increases. We shall see in Section 5.5 that successful systems can be obtained using a simple scaling scheme, involving the duplication of the rules grid (Sipper and Tomassini, 1996b). Below we report on a more sophisticated, empirically-obtained scheme, that has proven successful (Sipper et al., 1996c).

Given an evolved non-uniform CA of size N, our goal is to obtain a grid of size N', where N' is given but arbitrary (N' may be $> N$ or $< N$), such that the original performance level is maintained. This requires an algorithm for determining which rule should be placed in each cell of the size N' grid, so as to preserve the original grid's "essence," i.e., its emergent global behavior. Thus, we must determine what characterizes this latter behavior. We first note that there are two basic rule structures of importance in the original grid (shown for $r = 1$):

- The *local structure* with respect to cell i, $i \in \{0, \ldots, N-1\}$, is the set of three rules in cells $i-1$, i, and $i+1$ (indices are computed modulus N since the grid is circular).

- The *global structure* is derived by observing the blocks of identical rules present in the grid. For example, for the following evolved $N = 15$ grid:

| $R_1 R_1 R_1 R_1$ | $R_2 R_2$ | R_3 | $R_4 R_4 R_4 R_4$ | R_1 | $R_5 R_5 R_5$ |

  where R_j, $j \in \{1, \ldots, 5\}$, denotes a distinct rule, the number of blocks is 6, and the global structure is given by the list $\{R_1, R_2, R_3, R_4, R_1, R_5\}$.

We have found that if these structures are preserved, the scaled CA's behavior is identical to that of the original one. A heuristic principle is to expand (or reduce) a block of identical rules which spans at least four cells, while keeping intact blocks of length three or less. It is straightforward to observe that a block of length one or two should be left untouched, so as to maintain the local structure. As for a block of length three, there is no a-priori reason why it should be left unperturbed, rather, this has been found to hold empirically. A possible explanation may be that in such a three-cell block the local structure $R_j R_j R_j$ appears only once, thereby comprising a "primitive" unit that must be maintained. As an example of this procedure, consider the above $N = 15$ CA. Scaling this grid to size $N' = 25$ results in:

| $R_1 R_1 R_1 R_1 R_1 R_1 R_1 R_1$ | $R_2 R_2$ | R_3 | $R_4 R_4 R_4 R_4 R_4 R_4 R_4 R_4 R_4 R_4$ | R_1 | $R_5 R_5 R_5$ |

Note that both the local and global structures are preserved. We tested our scaling procedure on several CAs that were evolved to solve the synchronization task. The original grid sizes were $N = 100, 150$, which were then scaled to grids of sizes $N' = 200, 300, 350, 450, 500, 750$. In all cases the scaled grids exhibited the same (perfect) performance level as that of the original ones. An example of a scaled system is given in Figure 4.16.

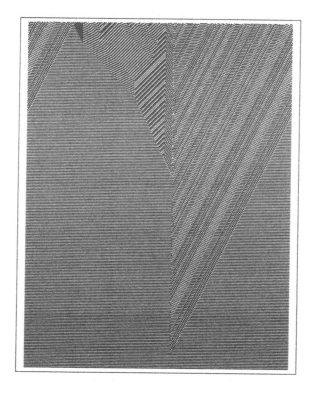

Figure 4.16. Example of a scaled CA: The size $N = 149$ CA of Figure 4.13b, evolved to solve the synchronization task, is scaled to $N' = 350$.

4.8 Discussion

In this chapter we described the cellular programming approach, used to evolve non-uniform CAs, and studied two non-trivial computational tasks, density and synchronization. The algorithm presented involves *local coevolution*, in contrast to the standard genetic algorithm in which independent problem solutions globally evolve. Our results demonstrate that: (1) non-uniform CAs can attain high computational performance on non-trivial problems, and (2) such systems can be evolved rather than designed. This is notable when one considers the huge search spaces involved, much larger than for uniform CAs.

Evolved, non-uniform CAs, with radius $r = 3$, attained high performance on the density task. We noted that rules are clustered in the vicinity of $\lambda = 0.5$, mostly in the region $0.45 - 0.55$. It is argued that these results lend support to the conclusion of Mitchell et al. (1993), stating that λ values for the density task are likely to be close to 0.5, rather than to critical λ_c values, as put forward by Packard (1988).

The small size of uniform, $r = 1$ CA rule space enabled us to test all possible rules on both tasks, finding for each one the uniform rule of maximal performance. We then demonstrated that non-uniform, $r = 1$ CAs can be evolved to perform these tasks with high performance, similar to uniform $r = 3$ CAs, and notably higher than uniform, $r = 1$ CAs. This suggests that non-uniformity reduces connectivity requirements, i.e., the use of smaller radiuses is made possible.

For $r = 1$, we observed that evolution tends to progress toward quasi-uniform grids. For the density task, type 2 quasi-uniformity was evolved, i.e., with one dominant rule. The crucial factor pertains to the fact that dominance is not total (in which case a uniform CA would result), rather a small number of other rules exists. The non-dominant rules act as buffers, preventing information from flowing too freely, while making local corrections to passing signals. A study of the genescape provided us with further insight into the evolutionary process, and a possible connection between the punctuated-equilibria phenomenon and the edge of chaos. For the synchronization task, type 1 quasi-uniformity was observed, with $2 - 3$ dominant rules. We noted that signals interact at the boundaries between rule blocks, thereby ultimately achieving synchronization. We showed that two-dimensional systems can be successfully evolved for both tasks, and presented a scheme for scaling evolved CAs.

Some qualitative differences have been detected between different values of r. The $r = 3$ case did not exhibit punctuated equilibria as with $r = 1$. For $r = 1$, all runs were successful, i.e., average fitness increased, as opposed to $r = 3$, where some runs were unsuccessful (e.g., Figure 4.3b). Quasi-uniform grids were not evolved for $r = 3$ as with $r = 1$. It cannot be ascertained at this point whether these are genuine differences or simply a result of our use of insufficient computational resources (grid size, number of configurations), however, there is some evidence pointing to the latter. For example, though quasi-uniformity was not observed for $r = 3$, we did note that the average Hamming distance between rules decreases as performance increases.[14]

A major impediment preventing ubiquitous computing with CAs stems from the difficulty of utilizing their complex behavior to perform useful computations. We noted in Section 4.2 that the GKL rule attains high performance on the density task. However, this is a serendipitous effect since the GKL rule was not invented for the purpose of performing any particular computational task (Mitchell et al., 1993). The difficulty of designing CAs to exhibit a specific behavior or perform a particular task has severely limited their applications; automating the design (programming) process would greatly enhance their viability. Our results offer encouraging prospects in this respect for non-uniform CAs. In the next chapter we study a number of novel computational tasks, for which cellular machines

---

[14]The Hamming distance between two rules is the number of bits by which their bit strings differ.

were evolved via cellular programming, suggesting possible application domains for our systems.

Chapter 5

Toward Applications of Cellular Programming

"Can you do Addition?" the White Queen asked. "What's one and one and one and one and one and one and one and one and one and one?"
"I don't know," said Alice. "I lost count."
"She can't do Addition," the Red Queen interrupted.

Lewis Carroll, *Through the Looking-Glass*

In the previous chapter we presented the cellular programming approach by which parallel cellular machines, namely, non-uniform cellular automata, are coevolved to perform computational tasks. In particular, we studied two non-trivial problems, density and synchronization, demonstrating that high-performance systems can be evolved to solve both. In this chapter we "revisit" the synchronization task, as well as study four novel ones, which are motivated by real-world applications. The complete list of tasks is given in Table 5.1.

Task	Description	Grid
Density	Decide whether the initial configuration contains a majority of 0s or of 1s	1D, r=1 2D, n=5
Synchronization	Given any initial configuration, relax to an oscillation between all 0s and all 1s	1D, r=1 2D, n=5
Ordering	Order initial configuration so that 0s are placed on the left and 1s are placed on the right	1D, r=1
Rectangle-Boundary	Find the boundaries of a randomly-placed, random-sized, non-filled rectangle	2D, n=5
Thinning	Find thin representations of rectangular patterns	2D, n=5
Random Number Generation	Generate "good" sequences of pseudo-random numbers	1D, r=1

Table 5.1. List of computational tasks for which parallel cellular machines were evolved via cellular programming. Right column delineates the grids used (one-dimensional, $r = 1$ and/or two-dimensional, with a 5-cell neighborhood).

5.1   The synchronization task revisited: Constructing counters

In the one-dimensional synchronization task, discussed in Section 4.5.2, the final pattern consists of an oscillation between all 0s and all 1s. From an engineering point of view, this period-2 cycle may be considered a 1-bit counter. Building upon such an evolved CA, using a small number of different cellular clock rates, 2- and 3-bit counters can be constructed (Marchal et al., 1997).

Constructing a 2-bit counter from a non-uniform, $r = 1$ CA, evolved to solve the synchronization task, is carried out by "interlacing" two $r = 1$ CAs, in the following manner: each cell in the evolved $r = 1$ CA is transformed into an $r = 2$ cell, two duplicates of which are juxtaposed (the resulting grid's size is thus doubled). This transformation is carried out by "blowing up" the $r = 1$ rule table into an $r = 2$ one, creating from each of the (eight) $r = 1$ table entries four $r = 2$ table entries, resulting in the 32-bit $r = 2$ rule table. For example, entry $110 \rightarrow 1$ specifies a next-state bit of 1 for an $r = 1$ neighborhood of 110 (left cell is in state 1, central cell is in state 1, right cell is in state 0). Transforming it into an $r = 2$ table entry is carried out by "moving" the adjacent, distance-1 cells to a distance of 2, i.e., $110 \rightarrow 1$ becomes $1X1Y0 \rightarrow 1$; filling in the four permutations of (X, Y), namely, $(0, 0)$, $(0, 1)$, $(1, 0)$, and $(1, 1)$, results in the four $r = 2$ table entries. The clocks of the odd-numbered cells function twice as fast as those of the even-numbered cells, meaning that the latter update their states every second time step with respect to the former. The resulting CA converges to a period-4 cycle upon presentation of a random initial configuration, a behavior that may be considered a 2-bit counter. The operation of such a CA is demonstrated in Figure 5.1.

Constructing a 3-bit counter from a non-uniform, $r = 1$ CA is carried out in a similar manner, by "interlacing" three $r = 1$ CAs (the resulting grid's size is thus tripled). The clocks of cells $0, 3, 6, \ldots$ function normally, those of cells $1, 4, 7, \ldots$ are divided by two (i.e., these cells change state every second time step with respect to the "fast" cells), and the clocks of cells $2, 5, 8, \ldots$ are divided by four (i.e., these cells change state every fourth time step with respect to the fast cells). The resulting CA converges to a period-8 cycle upon presentation of a random initial configuration, a behavior that may be considered a 3-bit counter. The operation of such a CA is shown in Figure 5.2. We have thus demonstrated how one can build upon an evolved behavior in order to construct devices of interest.

5.2   The ordering task

In this task, the non-uniform, one-dimensional, $r = 1$ CA, given any initial configuration, must reach a final configuration in which all 0s are placed on the left side of the grid and all 1s on the right side. This means that there are $N(1 - \rho(0))$ 0s on the left and $N\rho(0)$ 1s on the right, where $\rho(0)$ is the density of 1s at time 0,

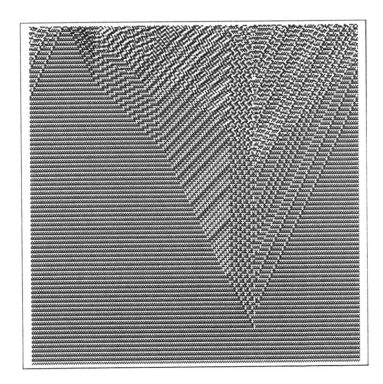

Figure 5.1. The one-dimensional synchronization task: A 2-bit counter. Operation of a non-uniform, 2-state CA, with connectivity radius $r = 2$, derived from the coevolved, $r = 1$ CA of Figure 4.13b. Grid size is $N = 298$ (twice that of the original $r = 1$ CA). The CA converges to a period-4 cycle upon presentation of a random initial configuration, a behavior that may be considered a 2-bit counter.

as defined in Section 4.2 (thus, the final density equals the initial one, however, the configuration consists of a block of 0s on the left, followed by a block of 1s on the right). The ordering task may be considered a variant of the density task and is clearly non-trivial following our reasoning of the previous chapter. It is interesting in that the output is not a uniform configuration of all 0s or all 1s, as with the density and synchronization tasks.

As with the previous studies of one-dimensional, $r = 1$ CAs, we tested all uniform, $r = 1$ CA rules on the ordering task. Results are depicted in Figure 5.3 (as explained in Section 4.5.1, rules are numbered in accordance with Wolfram's convention, Wolfram, 1983). We note that the highest fitness (performance) value is 0.71 attained by rule 232 (majority).

The cellular programming algorithm yielded non-uniform CAs with performance values similar to the density task, i.e., as high as 0.93. We also observed

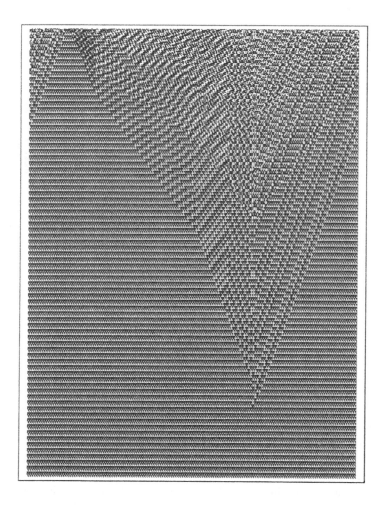

Figure 5.2. The one-dimensional synchronization task: A 3-bit counter. Operation of a non-uniform, 2-state CA, with connectivity radius $r = 3$, derived from the coevolved, $r = 1$ CA of Figure 4.13b. Grid size is $N = 447$ (three times that of the original $r = 1$ CA). The CA converges to a period-8 cycle upon presentation of a random initial configuration, a behavior that may be considered a 3-bit counter.

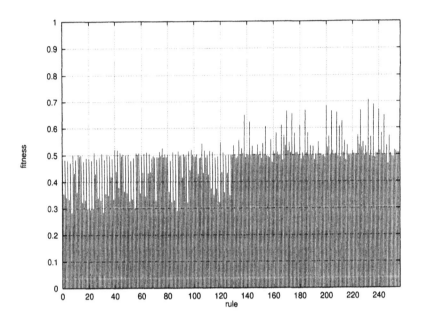

Figure 5.3. The ordering task: Fitness results of all possible uniform, $r = 1$ CA rules.

that evolved CAs are quasi-uniform, type-1, i.e., with a small number of dominant rules (Section 2.5). As with the density and synchronization tasks, we find that non-uniform CAs can be coevolved to attain high performance. While not perfect, it is notable that our coevolved, non-uniform, $r = 1$ CAs outperform *all* uniform, $r = 1$ ones. Figure 5.4 depicts the operation of a typical, coevolved CA.

The genescape of the non-uniform CA of Figure 5.4 is presented in Figure 5.5. We observe that genes 0 (000) and 7 (111) are the ones used most extensively, acting to preserve correct local information. We note that two genes have gained dominance, namely, genes 4 (100) and 6 (110). These genes are used when the local neighborhood ordering is incorrect, consisting of a 1 on the left and a 0 on the right. In such a situation signals are propagated, as evident in Figure 5.4, which explains the high usage of these genes.

5.3   The rectangle-boundary task

The possibility of applying CAs to perform image processing tasks arises as a natural consequence of their architecture. In a two-dimensional CA, a cell (or a group of cells) can correspond to an image pixel, with the CA's dynamics designed so as to perform a desired image processing task. Earlier work in this area, carried out mostly in the 1960s and the 1970s, was treated by Preston, Jr.

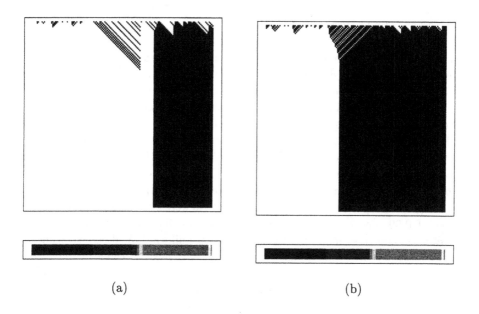

(a)   (b)

Figure 5.4. The ordering task: Operation of a coevolved, non-uniform, $r = 1$ CA. Grid size is $N = 149$. Top figures depict space-time diagrams, bottom figures depict rule maps. (a) Initial density of 1s is 0.315, final density is 0.328. (b) Initial density of 1s is 0.60, final density is 0.59.

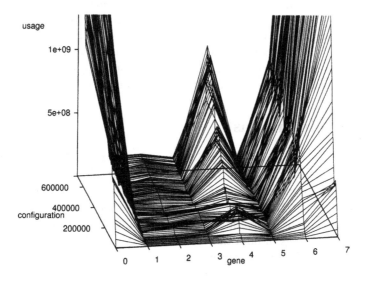

Figure 5.5. The ordering task: Genescape of a coevolved, non-uniform, $r = 1$ CA.

and Duff (1984), with more recent applications presented by Broggi et al. (1993) and Hernandez and Herrmann (1996).

The next two tasks involve image processing operations. In this section we discuss a two-dimensional boundary computation: given an initial configuration consisting of a non-filled rectangle, the CA must reach a final configuration in which the rectangular region is filled, i.e., all cells within the confines of the rectangle are in state 1, and all other cells are in state 0. Initial configurations consist of random-sized rectangles placed randomly on the grid (in our simulations, cells within the rectangle in the initial configuration were set to state 1 with probability 0.3; cells outside the rectangle were set to 0). Note that boundary cells can also be absent in the initial configuration. This operation can be considered a form of image enhancement, used, e.g., for treating corrupted images.

Using cellular programming, non-uniform CAs were evolved with performance values of 0.99, one of which is depicted in Figure 5.6. Figure 5.7 shows the two-dimensional rules map of the coevolved, non-uniform CA, demonstrating its quasi-uniformity, with one dominant rule occupying most of the grid. This rule maps the cell's state to zero if the number of neighboring cells in state 1 (including the cell itself) is less than two, otherwise mapping the cell's state to one.[1] Thus, growing regions of 1s are more likely to occur within the rectangle confines than without.

A one-dimensional version of the above problem is the line-boundary task, where the boundary consists of a (one-dimensional) line. Applying cellular programming yielded non-uniform, $r = 1$ CAs with peak performance of 0.94, one of which is depicted in Figure 5.8. We observed that a simple strategy had emerged, consisting of a quasi-uniform grid with two dominant rules, 252 and 238, whose tables are delineated in Figure 5.9. The rules differ in two positions, namely, the output bits of neighborhoods 100 and 001. Rule 252 maps neighborhood 100 to 1 and neighborhood 001 to 0, thus effecting a right-moving signal, while rule 238 analogously effects a left-moving signal. The line computation results from the interaction of these two rules (along with a small number of others).

5.4   The thinning task

Thinning (also known as skeletonization) is a fundamental preprocessing step in many image processing and pattern recognition algorithms. When the image consists of strokes or curves of varying thickness, it is usually desirable to reduce them to thin representations, located along the approximate middle of the original figure. Such "thinned" representations are typically easier to process in later stages, entailing savings in both time and storage space (Guo and Hall, 1989).

While the first thinning algorithms were designed for serial implementation, current interest lies in parallel systems, early examples of which were presented

---

[1]This is referred to as a totalistic rule, in which the state of a cell depends only on the sum of the states of its neighbors at the previous time step, and not on their individual states (Wolfram, 1983).

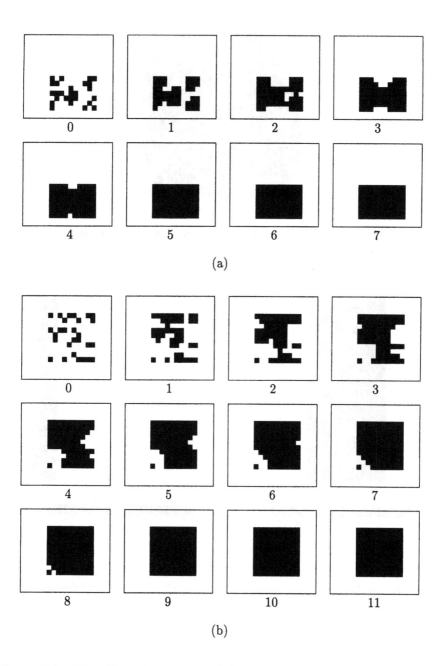

Figure 5.6. Two-dimensional rectangle-boundary task: Operation of a coe-volved, non-uniform, 2-state, 5-neighbor CA. Grid size is $N = 225$ (15×15). Numbers at bottom of images denote time steps. (a), (b) Computation is shown for two different initial configurations.

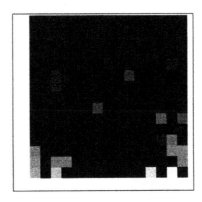

Figure 5.7. Two-dimensional rectangle-boundary task: Rules map of a coevolved, non-uniform, 2-state, 5-neighbor CA.

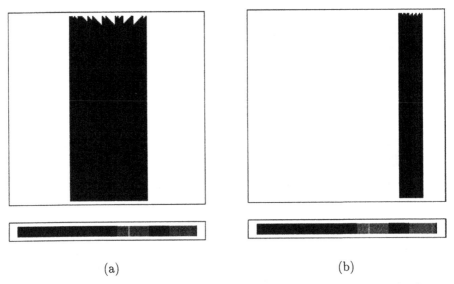

           (a)   (b)

Figure 5.8. One-dimensional line-boundary task: Operation of a coevolved, non-uniform, $r = 1$ CA. Top figures depict space-time diagrams, bottom figures depict rule maps. (a), (b) Computation is shown for two different initial configurations.

	Rule 252									Rule 238							
	7	6	5	4	3	2	1	0		7	6	5	4	3	2	1	0
neighborhood	111	110	101	100	011	010	001	000		111	110	101	100	011	010	001	000
output bit	1	1	1	1	1	1	0	0		1	1	1	0	1	1	1	0

Figure 5.9. Rules involved in the line-boundary task.

by Preston, Jr. and Duff (1984). The difficulty of designing a good thinning algorithm using a small, local cellular neighborhood, coupled with the task's importance, has motivated us to explore the possibility of applying cellular programming.

Guo and Hall (1989) considered four sets of binary images, two of which consist of rectangular patterns oriented at different angles. The algorithms presented therein employ a two-dimensional grid with a 9-cell neighborhood; each parallel step consists of two sub-iterations in which distinct operations take place. The set of images considered by us consists of rectangular patterns oriented either horizontally or vertically. While more restrictive than that of Guo and Hall (1989), it is noted that we employ a smaller neighborhood (5-cell) and do not apply any sub-iterations.

Figure 5.10 demonstrates the operation of a coevolved CA performing the thinning task. Although the evolved grid does not compute perfect solutions, we observe, nonetheless, good thinning "behavior" upon presentation of rectangular patterns as defined above (Figure 5.10a). Furthermore, partial success is demonstrated when presented with more difficult images, involving intersecting lines (Figure 5.10b).

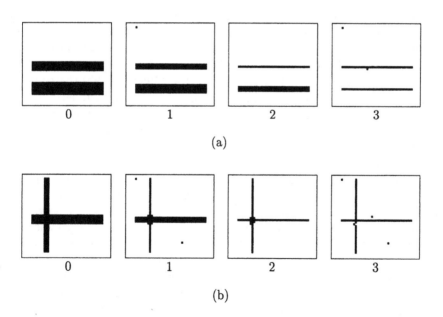

Figure 5.10. Two-dimensional thinning task: Operation of a coevolved, non-uniform, 2-state, 5-neighbor CA. Grid size is $N = 1600$ (40×40). Numbers at bottom of images denote time steps. (a) Two separate lines. (b) Two intersecting lines.

5.5   Random number generation

Random numbers are needed in a variety of applications, yet finding good random number generators, or randomizers, is a difficult task (Park and Miller, 1988). To generate a random sequence on a digital computer, one starts with a fixed-length seed, then iteratively applies some transformation to it, progressively extracting as long as possible a random sequence. Such numbers are usually referred to as *pseudo*-random, as distinguished from true random numbers, resulting from some natural physical process. In order to demonstrate the efficacy of a proposed generator, it is usually subject to a battery of empirical and theoretical tests, among which the most well known are those described by Knuth (1981).

In the last decade cellular automata have been used to generate "good" random numbers. The first work examining the application of CAs to random number generation is that of Wolfram (1986), in which the uniform, 2-state, $r = 1$ rule-30 CA was extensively studied, demonstrating its ability to produce highly random, temporal bit sequences. Such sequences are obtained by sampling the values that a particular cell (usually the central one) attains as a function of time.

In Wolfram's work, the uniform rule-30 CA is initialized with a configuration consisting of a single cell in state 1, with all other cells being in state 0 (Wolfram, 1986). The initially non-zero cell is the site at which the random temporal sequence is generated. Wolfram studied this particular rule extensively, demonstrating its suitability as a high-performance randomizer, which can be efficiently implemented in parallel; indeed, this CA is one of the standard generators of the massively parallel Connection Machine CM2 (Connection, 1991). A non-uniform CA randomizer was presented by Hortensius et al. (1989a; 1989b) (based on the work of Pries et al., 1986), consisting of two rules, 90 and 150, arranged in a specific order in the grid. The performance of this CA in terms of random number generation was found to be at least as good as that of rule 30, with the added benefit of less costly hardware implementation. It is interesting in that it combines two rules, both of which are simple linear rules, that do not comprise good randomizers, to form an efficient, high-performance generator. An example application of such CA randomizers was demonstrated by Chowdhury et al. (1995) who presented the design of a low-cost, high-capacity associative memory.

An evolutionary approach for obtaining random number generators was taken by Koza (1992), who applied genetic programming to the evolution of a symbolic LISP expression that acts as a rule for a uniform CA (i.e., the expression is inserted into each CA cell, thereby comprising the function according to which the cell's next state is computed). He demonstrated evolved expressions that are equivalent to Wolfram's rule 30. The fitness measure used by Koza is the *entropy* E_h: let k be the number of possible values per sequence position (in our case CA states) and h a subsequence length. E_h (measured in bits) for the set of k^h probabilities of the k^h possible subsequences of length h is given by:

$$E_h = - \sum_{j=1}^{k^h} p_{h_j} \log_2 p_{h_j}$$

where $h_1, h_2, \ldots, h_{k^h}$ are all the possible subsequences of length h (by convention, $\log_2 0 = 0$ when computing entropy). The entropy attains its maximal value when the probabilities of all k^h possible subsequences of length h are equal to $1/k^h$; in our case $k = 2$ and the maximal entropy is $E_h = h$. Koza evolved LISP expressions which act as rules for uniform, one-dimensional CAs. The CAs were run for 4096 time steps and the entropy of the resulting temporal sequence of a designated cell (usually the central one) was taken as the fitness of the particular rule (i.e., LISP expression). In his experiments, Koza used a subsequence length of $h = 4$, obtaining rules with an entropy of 3.996. The best rule of each run was re-tested over 65536 time steps, some of which exhibited the maximal entropy value of 4.0.

The above account leads us to ask whether good CA randomizers can be coevolved using cellular programming; the results reported below suggest that indeed this is the case (Sipper and Tomassini, 1996a; Sipper and Tomassini, 1996b). The algorithm presented in Section 4.3 is slightly modified, such that the cell's fitness score for a single configuration is defined as the entropy E_h of the temporal sequence, after the CA has been run for M time steps. Cell i's fitness value, f_i, is then updated as follows (refer to Figure 4.2):

for each cell i **do in parallel**
$\qquad f_i = f_i +$ entropy E_h of the temporal sequence of cell i
end parallel for

Initial configurations are randomly generated and for each one the CA is run for $M = 4096$ time steps.[2] Note that we do not restrict ourselves to one designated cell, but consider all grid cells, thus obtaining N random sequences in parallel, rather than a single one.

In our simulations (using grids of sizes $N = 50$ and $N = 150$), we observed that the average cellular entropy taken over all grid cells is initially low (usually in the range $[0.2, 0.5]$), ultimately evolving to a maximum of 3.997, when using a subsequence size of $h = 4$ (i.e., entropy is computed by considering the oc-currence probabilities of 16 possible subsequences, using a "sliding window" of length 4). The progression of a typical evolutionary run is depicted in Figure 5.11. We performed several such experiments using $h = 4$ and $h = 7$. The evolved, non-uniform CAs attained average fitness values (entropy) of 3.997 and 6.978, respectively. We then re-tested our best CAs over $M = 65536$ times steps (as in Koza, 1992), obtaining entropy values of 3.9998 and 6.999, respectively. Interest-ingly, when we performed this test with $h = 7$ for CAs which were evolved using $h = 4$, high entropy was displayed, as for CAs which were originally evolved with $h = 7$. These results are comparable to the entropy values obtained by Koza (1992), as well as to those of the rule-30 CA of Wolfram (1986) and the non-uniform, rules $\{90, 150\}$ CA of Hortensius et al. (1989a; 1989b). Note that while our fitness measure is local, the evolved entropy results reported above represent the average of *all* grid cells. Thus, we obtain N random sequences in parallel

---

[2]A standard, 48-bit, linear congruential algorithm proved sufficient for the generation of random initial configurations.

rather than a single one. Figure 5.12 demonstrates the operation of three CAs discussed above: rule 30, rules $\{90, 150\}$, and a coevolved CA. Note that the latter is quasi-uniform, type 1, as evident by observing the rules map.

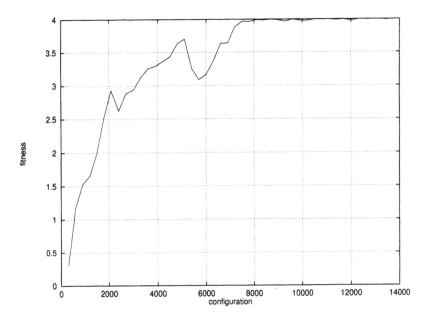

Figure 5.11. One-dimensional random number generator ($r = 1$): Progression of a typical evolutionary run. Graph depicts the average fitness of all grid cells, as a function of the number of configurations presented so far. Cellular fitness f_i equals entropy E_h (shown for $h = 4$).

We next subjected our evolved CAs to a number of additional tests, including chi-square (χ^2), serial correlation coefficient, and a Monte Carlo simulation for calculating the value of π; these are well-known tests described in detail by Knuth (1981). In order to apply the tests we generated sequences of $100,000$ random bytes using two different procedures: (a) The CA of size $N = 50$ is run for 500 time steps, thus generating 50 random temporal bit sequences of length 500. These are concatenated to form one long sequence of length $25,000$ bits. This process is then repeated 32 times, thus obtaining a sequence of $800,000$ bits, which are grouped into $100,000$ bytes. (b) The CA of size $N = 50$ is run for 400 time steps. Every 8 time steps, 50 8-bit sequences (bytes) are produced, which are concatenated, resulting in 2500 bytes after 400 time steps. This process is then repeated 40 times, thus obtaining the $100,000$ byte sequence.

Table 5.2 shows the test results of four random number generators:[3] two

---

[3]The tests were conducted using a public-domain software written by J. Walker, available at http://www.fourmilab.ch/random/.

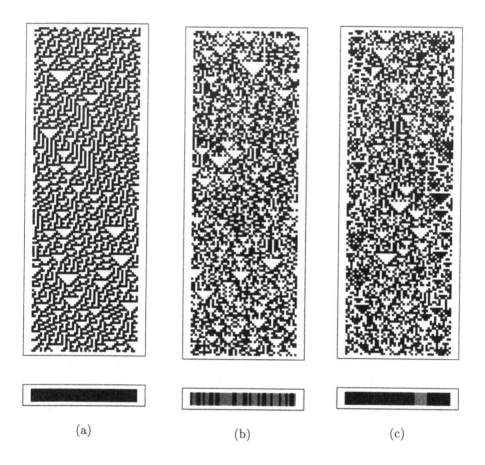

Figure 5.12. One-dimensional random number generators: Operation of three CAs. Grid size is $N = 50$, radius is $r = 1$. Initial configurations were generated by randomly setting the state of each grid cell to 0 or 1 with uniform probability. Top figures depict space-time diagrams, bottom figures depict rule maps. (a) Rule-30 CA. (b) Rules $\{90, 150\}$ CA. (c) A coevolved, non-uniform CA, consisting of three rules: rule 165 (22 cells), rule 90 (22 cells), and rule 150 (6 cells).

Test	coevolved CA (1)		coevolved CA (2)		rule 30 CA		rules {90,150} CA	
	(a)	(b)	(a)	(b)	(a)	(b)	(a)	(b)
	50.00%	75.00%	50.00%	50.00%	90.00%	90.00%	50.00%	25.00%
	50.00%	50.00%	75.00%	50.00%	10.00%	50.00%	5.00%	50.00%
	90.00%	50.00%	95.00%	5.00%	97.50%	0.50%	10.00%	50.00%
	25.00%	75.00%	50.00%	50.00%	0.01%	50.00%	75.00%	25.00%
i	50.00%	25.00%	75.00%	50.00%	95.00%	75.00%	97.50%	25.00%
	25.00%	10.00%	75.00%	25.00%	97.50%	50.00%	25.00%	50.00%
	75.00%	50.00%	75.00%	75.00%	50.00%	50.00%	25.00%	50.00%
	10.00%	50.00%	25.00%	50.00%	5.00%	50.00%	25.00%	50.00%
	50.00%	25.00%	50.00%	75.00%	25.00%	50.00%	95.00%	75.00%
	90.00%	75.00%	90.00%	10.00%	25.00%	50.00%	75.00%	75.00%
	100%	100%	90%	90%	50%	90%	70%	100%
	0.00185	-0.00085	-0.00390	0.01952	0.00052	-0.24685	0.00646	0.00036
	-0.00386	-0.00228	0.00228	0.02144	-0.00175	-0.24838	-0.00071	-0.00194
ii	0.00192	-0.00297	0.00048	0.01970	0.00156	-0.24291	0.00205	-0.00322
	-0.00011	-0.00248	-0.00237	0.02192	0.00478	-0.23735	0.00177	0.00094
	-0.00060	-0.00762	0.00194	0.01937	0.00214	-0.24149	-0.00075	0.00378
	7.99819	7.99828	7.99807	7.99827	7.99841	7.99842	7.99821	7.99797
	7.99821	7.99817	7.99835	7.99810	7.99789	7.99820	7.99788	7.99807
iii	7.99838	7.99810	7.99845	7.99786	7.99849	7.99770	7.99793	7.99809
	7.99800	7.99831	7.99806	7.99808	7.99733	7.99807	7.99832	7.99804
	7.99808	7.99801	7.99829	7.99808	7.99844	7.99835	7.99851	7.99800
	0.54%	0.19%	0.42%	0.16%	0.21%	0.90%	0.52%	0.20%
iv	0.03%	0.12%	0.33%	0.35%	0.21%	0.13%	0.05%	0.07%
	0.18%	0.68%	0.62%	0.65%	0.32%	0.13%	0.27%	0.07%
	0.45%	0.73%	0.48%	0.33%	0.37%	0.38%	0.07%	0.17%
	0.16%	0.09%	0.12%	0.13%	0.40%	0.08%	0.78%	0.01%

Table 5.2. Results of tests. Each entry represents the test result for a sequence of 100,000 bytes, generated by the corresponding randomizer. 20 sequences were generated by each randomizer, 10 by procedure (a) and 10 by procedure (b) (see text). The table lists the chi-square test results for all 10 sequences and the first 5 results for the other tests. CA Grid size is $N = 50$. Coevolved CA (1) consists of three rules: rule 165 (22 cells), rule 90 (22 cells), and rule 150 (6 cells). Coevolved CA (2) consists of two rules: rule 165 (45 cells) and rule 225 (5 cells). Interpretation of the listed values is as follows (for a full explanation see Knuth, 1981): (i) Chi-square test: "good" results are between $10\% - 90\%$, with extremities on both sides (i.e., $< 10\%$ and $> 90\%$) representing non-satisfactory random sequences. The total percentage of sequences passing the chi-square test is listed below the 10 individual test results. Knuth suggests that at least three sequences from a generator be subject to the chi-square test and if a majority pass then the generator is considered to have passed (with respect to chi-square). (ii) Serial correlation coefficient: this value should be close to zero. (iii) Entropy test: this value should be close to 8. (iv) Monte Carlo π: the random number sequence is used in a Monte Carlo computation of the value of π, and the error percentage from the actual value is shown.

CA	% success	Chi-square test results										
(1)	100%	(a)	75%	50%	25%	25%	25%	75%	50%	50%	90%	50%
	90%	(b)	75%	50%	50%	97.5%	50%	75%	50%	50%	75%	50%
(2)	80%	(a)	50%	50%	25%	50%	2.5%	50%	95%	50%	50%	50%
	90%	(b)	50%	50%	50%	75%	25%	50%	50%	25%	5%	10%

Table 5.3. Chi-square test results of two scaled, $N = 500$ CAs. These were created from the corresponding coevolved, $N = 50$ CAs (CAs (1) and (2)), by duplicating the evolved grid ten times. 20 random sequences were generated by each CA, 10 by procedure (a) and 10 by procedure (b). Chi-square test results are shown, along with the percentage of sequences passing the test. The other tests were also conducted, obtaining similar results to the original, non-scaled CAs.

coevolved CAs, rule-30 CA, and the rules $\{90, 150\}$ CA. We note that the two coevolved CAs attain good results on all tests, most notably chi-square which is one of the most significant ones (Knuth, 1981). Our results are somewhat better than the rules $\{90, 150\}$ CA, and markedly improved in comparison to the rule-30 CA, which attains lower scores on the chi-square test (procedure (a)), and on the serial correlation test (procedure (b)). It is noteworthy that our CAs attain good results on a number of tests, while the fitness measure used during evolution is entropy alone. The relatively low results obtained by the rule-30 CA may be due to the fact that we considered N random sequences generated in parallel, rather than the single one considered by Wolfram. We note in passing that the rules $\{90,150\}$ CA results may probably be somewhat improved (as perhaps our own results) by using "site spacing" and "time spacing" (Hortensius et al., 1989a; Hortensius et al., 1989b). Our final experiment involves the implementation of a simple scaling scheme, in which a $N = 500$ CA is created from the evolved $N = 50$ one. This is done in a straightforward manner by duplicating the evolved CA 10 times, i.e., concatenating 10 identical copies of the 50-cell rules grid. While simpler than the scaling procedure described in Section 4.7, Table 5.3 shows that good randomizers can thus be obtained.

We have shown that the cellular programming algorithm can be applied to the difficult problem of generating random number generators. While a more extensive suite of tests is in order, it seems safe to say at this point that our coevolved generators are at least as good as the best available CA randomizers. Furthermore, there is a notable advantage arising from the existence of a "tunable" algorithm for the generation of randomizers.

We observed that our evolved CAs are quasi-uniform, involving a small number of distinct rules. As some rules lend themselves more easily to hardware implementation, our algorithm may be used to find good randomizers which can also be efficiently implemented. A possible extension is the addition of restrictions to the evolutionary process, e.g., by prespecifying rules for some cells, in order to accommodate hardware constraints. Another possible modification of the evolutionary process is the incorporation of statistical measures of randomness into the fitness function (and not just as an aftermath benchmark). These possible extensions could lead to the automatic generation of high-performance, random number generators, meeting specific user demands.

5.6   Concluding remarks

In this chapter we studied a number of computational tasks, motivated by real-world applications. We demonstrated that non-uniform CAs can evolve to perform these tasks with high performance. In the next chapter we present the "firefly" machine, an evolving, online, autonomous hardware system that implements the cellular programming algorithm, thus demonstrating that *evolware* can be attained.

Chapter 6

Online Autonomous Evolware: The Firefly Machine

Glories, like glow-worms, afar off shine bright,
But look'd too near have neither heat nor light.
John Webster, *The White Devil, Act iv, Scene 4*

6.1 Introduction

In this chapter we describe a hardware implementation of the cellular programming algorithm, thus demonstrating that "evolving ware," *evolware*, can be attained (Goeke et al., 1997; Sipper, 1997a; Sipper, 1996b). The underlying motivation for constructing the machine described herein was to demonstrate that *online* evolution can be attained, which operates without any reference to an external computer (see below). Toward this end we concentrated on a specific, well-defined problem, for which high performance can be attained, our choice being the one-dimensional synchronization task (Section 4.5.2). As a reminder, in this task the $r = 1$ CA, given any initial configuration, must reach a final configuration, within a prespecified number of time steps, that oscillates between all 0s and all 1s on successive time steps. Appropriately, the machine has been dubbed "firefly."[1]

The firefly project is part of an ongoing effort within the burgeoning field of bio-inspired systems and evolvable hardware (Sanchez and Tomassini, 1996). Most work carried out to date under this heading involves the application of evolutionary algorithms to the synthesis of digital systems (recently, analog systems have been studied as well, see Koza et al., 1996). From this perspective, evolvable hardware is simply a sub-domain of artificial evolution, where the final goal is the synthesis of an electronic circuit (Sanchez et al., 1997). However, several researchers have set more far-reaching goals for the field as a whole.

---

[1] See Section 4.2 for the relationship between the synchronization problem and fireflies in nature.

Current and (possible) future evolving hardware systems can be classified according to two distinguishing characteristics. The first involves the distinction between *offline* genetic operations, carried out in software, and *online* ones, which take place on an actual circuit. The second characteristic concerns *open-endedness*. When the fitness criterion is imposed by the user in accordance with the task to be solved (currently the rule with artificial-evolution techniques), one attains a form of *guided* evolution. This is to be contrasted with *open-ended* evolution occurring in nature, which admits no externally-imposed fitness criterion, but rather an implicit, emergent, dynamical one (that could arguably be summed up as "survivability"). In view of these two characteristics, one can define the following four categories of evolvable hardware (Sanchez et al., 1997):

- The first category can be described as *evolutionary circuit design*, where the entire evolutionary process takes place in software, with the resulting solution possibly loaded onto a real circuit. Though a potentially useful design methodology, this falls completely within the realm of traditional evolutionary techniques, as noted above. As examples one can cite the works of Koza et al. (1996), Hemmi et al. (1996), Kitano (1996), and Higuchi et al. (1996).

- The second category involves systems in which a real circuit is used during the evolutionary process, though most operations are still carried out offline, in software. As examples one can cite Murakawa et al. (1996), Iwata et al. (1996), Thompson et al. (1996), and Thompson (1997), where fitness calculation is carried out on a real circuit.

- In the third category one finds systems in which *all* genetic operations (selection, crossover, mutation, and fitness evaluation) are carried out *online*, in hardware. The major aspect missing concerns the fact that evolution is not open ended, i.e., there is a predefined goal and no dynamic environment to speak of. An example is the firefly machine described herein (Goeke et al., 1997).

- The last category involves a *population* of hardware entities evolving in an *open-ended* environment.

It has been argued that only systems within the last category can be truly considered evolvable hardware,[2] a goal which still eludes us at present (Sanchez et al., 1997). A natural application area for such systems is within the field of autonomous robots, which involves machines capable of operating in unknown environments without human intervention (Brooks, 1991). A related application domain is that of controllers for noisy, changing environments. Another interesting example would be what has been dubbed by Sanchez et al. (1997) "Hard-Tierra." This involves the hardware implementation of the Tierra "world," which consists of an open-ended environment of evolving computer programs (Ray, 1992;

---

[2]A more correct term would probably be *evolving* hardware.

see also Section 3.5). A small-scale experiment along this line was undertaken by Galley and Sanchez (1996). The idea of Hard-Tierra is interesting since it leads us to the observation that 'open-endedness' does not necessarily imply a real, biological environment. The firefly machine, belonging to the third category, demonstrates that complete online evolution can be attained, though not in an open-ended environment. This latter goal remains a prime target for future research.

In Section 6.2 we briefly present large-scale programmable circuits, specifically concentrating on Field-Programmable Gate Arrays (FPGA). An FPGA can be programmed "on the fly," thus offering an attractive technological platform for realizing, among others, evolware. In Section 6.3 we describe the FPGA-based firefly machine. Evolution takes place on-board, with no reference to or aid from any external device (such as a computer that carries out genetic operators), thus attaining online autonomous evolware. Finally, some concluding remarks are presented in Section 6.4.

6.2   Large-scale programmable circuits

An integrated circuit is called programmable when the user can configure its function by programming. The circuit is delivered after manufacturing in a generic state and the user can adapt it by programming a particular function. The programmed function is coded as a string of bits, representing the configuration of the circuit. In this chapter we consider solely programmable *logic* circuits, where the programmable function is a logic one, ranging from simple boolean functions to complex state machines.

The first programmable circuits allowed the implementation of logic circuits that were expressed as a logic sum of products. These are the PLDs (Programmable Logic Devices), whose most popular version is the PAL (Programmable Array Logic). More recently, a novel technology has emerged, affording higher flexibility and more complex functionality: the Field-Programmable Gate Array, or FPGA (Sanchez, 1996). An FPGA is an array of logic cells placed in an infrastructure of interconnections, which can be programmed at three distinct levels (Figure 6.1): (1) the function of the logic cells, (2) the interconnections between cells, and (3) the inputs and outputs. All three levels are programmed via a string of bits that is loaded from an external source, either once or several times. In the latter case the FPGA is considered *reconfigurable*.

FPGAs are highly versatile devices that offer the designer a wide range of design choices. However, this potential power necessitates a plethora of tools in order to design a system. Essentially, these generate the configuration bit string upon given such inputs as a logic diagram or a high-level functional description.

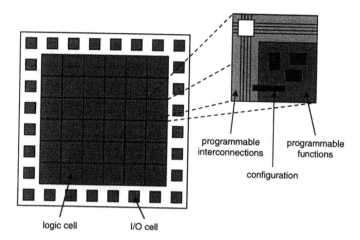

Figure 6.1. A schematic diagram of a Field-Programmable Gate Array (FPGA). An FPGA is an array of logic cells placed in an infrastructure of interconnections, which can be programmed at three distinct levels: (1) the function of the logic cells, (2) the interconnections between cells, and (3) the inputs and outputs. All three levels are programmed via a configuration bit string that is loaded from an external source, either once or several times.

6.3   Implementing evolware

In this section we describe the firefly evolware machine, an online implementation of the cellular programming algorithm (Section 4.3). To facilitate implementation, the algorithm is slightly modified (with no loss in performance): the two genetic operators, one-point crossover and mutation, are replaced by a single operator, *uniform crossover*. Under this operation, a new rule, i.e., an "offspring" genome, is created from two "parent" genomes (bit strings) by choosing each offspring bit from one or the other parent, with a 50% probability for each parent (Mitchell, 1996; Tomassini, 1996). The changes to the algorithm are therefore as follows (refer to Figure 4.2):

   else if $nf_i(c) = 1$ **then** replace rule i with the fitter neighboring rule,
         without mutation
   else if $nf_i(c) = 2$ **then** replace rule i with the *uniform* crossover of the
         two fitter neighboring rules, *without mutation*

The evolutionary process ends following an arbitrary decision by an outside observer (the '**while** not done' loop of Figure 4.2).

   The cellular programming evolware is implemented on a physical board whose only link to the "external world" is the power-supply cable. The features distinguishing this implementation from previous ones (described in Sanchez and

Tomassini, 1996) are: (1) an ensemble of individuals (cells) is at work rather than a single one, (2) genetic operators are all carried out on-board, rather than on a remote, offline computer, and (3) the evolutionary phase does not necessitate halting the machine's operation, but is rather intertwined with normal execution mode. These features entail an online autonomous evolutionary process.

The active components of the evolware board comprise exclusively FPGA circuits, with no other commercial processor whatsoever. An LCD screen enables the display of information pertaining to the evolutionary process, including the current rule and fitness value of each cell. The parameters M (number of time steps a configuration is run) and C (number of configurations between evolutionary phases, see Section 4.3) are tunable through on-board knob selectors; in addition, their current values are displayed. The implemented grid size is $N = 56$ cells, each of which includes, apart from the logic component, a LED indicating its current state (on=1, off=0), and a switch by which its state can be manually set.[3] We have also implemented an on-board global synchronization detector circuit, for the sole purpose of facilitating the external observer's task; this circuit is *not* used by the CA in any of its operational phases. The machine is depicted in Figure 6.2.

The architecture of a single cell is shown in Figure 6.3. The binary state is stored in a D-type flip-flop whose next state is determined either randomly, enabling the presentation of random initial configurations, or by the cell's rule table, in accordance with the current neighborhood of states. Each bit of the rule's bit string is stored in a D-type flip-flop whose inputs are channeled through a set of multiplexors, according to the current operational phase of the system:

1. During the initialization phase of the evolutionary algorithm, the (eight) rule bits are loaded with random values. This is carried out once per evolutionary run.

2. During the execution phase of the CA, the rule bits remain unchanged. This phase lasts a total of $C * M$ time steps (C configurations, each one run for M time steps).

3. During the evolutionary phase, and depending on the number of fitter neighbors, $nf_i(c)$ (Section 4.3), the rule is either left unchanged ($nf_i(c) = 0$), replaced by the fitter left or right neighboring rule ($nf_i(c) = 1$), or replaced by the uniform crossover of the two fitter rules ($nf_i(c) = 2$).

The (local) fitness score for the synchronization task is assigned to each cell by considering the last four time steps (i.e., $[M + 1..M + 4]$). If the sequence of states over these steps is precisely $0 \rightarrow 1 \rightarrow 0 \rightarrow 1$ (i.e., an alternation of 0s and 1s, starting from 0), the cell's fitness score is 1, otherwise this score is 0.

To determine the cell's fitness score for a single initial configuration, i.e., after the CA has been run for $M + 4$ time steps, a four-bit shift register is used

---

[3]This is used to test the evolved system after termination of the evolutionary process, by manually loading initial configurations.

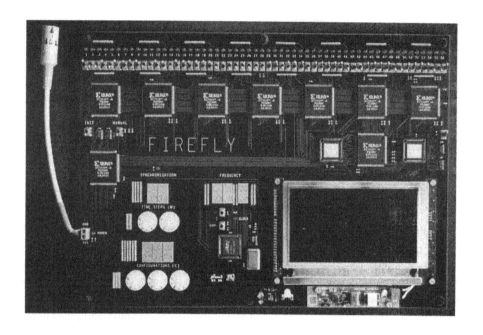

Figure 6.2. The firefly evolware machine. The system is a one-dimensional, non-uniform, $r = 1$ cellular automaton that evolves via execution of the cellular programming algorithm. Each of the 56 cells contains the genome representing its rule table; these genomes are randomly initialized, after which evolution takes place. The board contains the following components: (1) LED indicators of cell states (top), (2) switches for manually setting the initial states of cells (top, below LEDs), (3) Xilinx FPGA chips (below switches), (4) display and knobs for controlling two parameters ('time steps' and 'configurations') of the cellular programming algorithm (bottom left), (5) a synchronization indicator (middle left), (6) a clock pulse generator with a manually-adjustable frequency from 0.1 Hz to 1 MHz (bottom middle), (7) an LCD display of evolved rule tables and fitness values obtained during evolution (bottom right), and (8) a power-supply cable (extreme left). (Note that this is the system's sole external connection.)

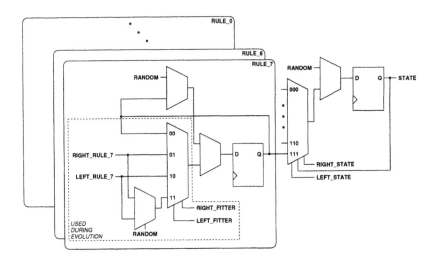

Figure 6.3. Circuit design of a cell. The binary state is stored in a D-type flip-flop whose next state is determined either randomly, enabling the presentation of random initial configurations, or by the cell's rule table, in accordance with the current neighborhood of states. Each bit of the rule's bit string is stored in a D-type flip-flop whose inputs are channeled through a set of multiplexors, according to the current operational phase of the system (initialization, execution, or evolution).

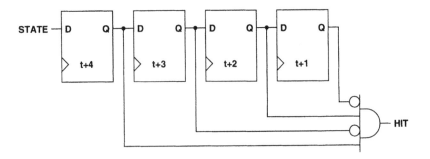

Figure 6.4. Circuit used (in each cell) after execution of an initial configuration to detect whether a cell receives a fitness score of 1 (HIT) or 0 (no HIT).

(Figure 6.4). This register continuously stores the states of the cell over the last four time steps ($[t+1..t+4]$). An AND gate tests for occurrence of the "good" final sequence (i.e., $0 \rightarrow 1 \rightarrow 0 \rightarrow 1$), producing the HIT signal, signifying whether the fitness score is 1 (HIT) or 0 (no HIT).

Each cell includes a fitness counter and two comparators for comparing the cell's fitness value with those of its two neighbors. Note that the cellular connections are entirely local, a characteristic enabled by the local operation of the cellular programming algorithm. In the interest of cost reduction, a number of resources have been implemented within a central control unit, including the random number generator and the M and C counters. These are implemented on-board and do not comprise a breach in the machine's autonomous mode of operation.

The random number generator is implemented with a linear feedback shift register (LFSR), producing a random bit stream that cycles through $2^{32} - 1$ different values (the value 0 is excluded since it comprises an undesirable attractor). As a cell uses at most eight different random values at any given moment, it includes an 8-bit shift register through which the random bit stream propagates. The shift registers of all grid cells are concatenated to form one large stream of random bit values, propagating through the entire CA. Cyclic behavior is eschewed due to the odd number of possible values produced by the random number generator ($2^{32} - 1$) and to the even number of random bits per cell.

6.4   Concluding remarks

We described an FPGA-based implementation of the cellular programming algorithm, the firefly machine, that exhibits complete online evolution, all genetic operators carried out in hardware, with no reference to an external computer. Firefly thus belongs to the third category of evolving hardware, described in Section 6.1. The major aspect missing concerns the fact that evolution is not open ended, i.e., there is a predefined goal and no dynamic environment to speak of. Open-endedness remains a prime target for future research in the field. We note that the machine's construction was facilitated by the cellular programming algorithm's local dynamics, highlighting a major advantage of such local evolutionary processes.

Evolware systems such as firefly enable enormous gains in execution speed. The cellular programming algorithm, when run on a high-performance workstation, executes 60 initial configurations per second.[4] In comparison, the firefly machine executes 13,000 initial configurations per second.[5]

The evolware machine was implemented using FPGA circuits, configured such that each cell within the system behaves in a certain general manner, after which

---

[4]This was measured using a grid of size $N = 56$, each initial configuration being run for $M = 75$ time steps, with the number of configurations between evolutionary phases $C = 300$.

[5]This is achieved when the machine operates at the current maximal frequency of 1 MHz. In fact, this can easily be increased to 6 MHz, thereby attaining 78,000 configurations per second.

evolution is used to "find" the cell's specific behavior, i.e., its rule table. Thus, the system consists of a fixed part and an evolving part, both specified via FPGA configuration strings (Figure 6.5). An interesting outlook on this setup is to consider the evolutionary process as one where an organism evolves within a given species, the former specified by the FPGA's evolving part, the latter specified by the fixed part. This raises the interesting issue of evolving the species itself.

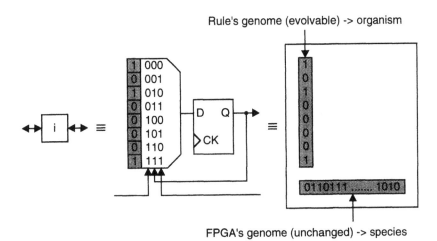

Figure 6.5. The firefly cell is hierarchically organized, consisting of two parts: (1) the "organismic" level, comprising an evolving configuration string that specifies the cell's rule table, and (2) the "species" level, a fixed (non-evolved) configuration string that defines the underlying FPGA's behavior.

Chapter 7

Studying Fault Tolerance in Evolved Cellular Machines

Further investigation quickly established what it was that had happened. A meteorite had knocked a large hole in the ship. The ship had not previously detected this because the meteorite had neatly knocked out that part of the ship's processing equipment which was supposed to detect if the ship had been hit by a meteorite.

Douglas Adams, *Mostly Harmless*

7.1 Introduction

Most classical software and hardware systems, especially parallel ones, exhibit a very low level of fault tolerance, i.e., they are not resilient in the face of errors; indeed, where software is concerned, even a single error can often bring an entire program to a grinding halt. Future computing systems may contain thousands or even millions of computing elements (e.g., Drexler, 1992). For such large numbers of components, the issue of resilience can no longer be ignored, since faults will be likely to occur with high probability.

Networks of automata exhibit a certain degree of fault tolerance. As an example, one can cite artificial neural networks, many of which show graceful degradation in performance when presented with noisy input. Moreover, the malfunction of a neuron or damage to a synaptic weight causes but a small change in the system's overall behavior, rather than bringing it to a complete standstill. Cellular computing systems, such as CAs, may be regarded as a simple and convenient framework within which to study the effects of such errors. Another motivation for studying this issue derives directly from the work presented in the previous chapter concerning the firefly machine. We wish to learn how robust such a machine is when operating under faulty conditions (Sipper et al., 1996a; Sipper et al., 1996b).

In this chapter we study the effects of random faults on the behavior of one-dimensional CAs obtained via cellular programming. In particular, we are interested in the systems' behavior as a function of the error level. We wish to learn whether there exist error-rate regions in which the automata can be considered to perform their task in an "acceptable" manner. Moreover, the amount and speed of *recovery* after the appearance of a fault is quantified and measured. We also observe how disturbances spread throughout the system to learn under what conditions the perturbation remains limited and does not propagate to the entire system.

In the next section related fault studies in cellular systems are briefly reviewed, followed by Section 7.3, describing probabilistic faults in CAs, as well as the "system replicas" framework within which to study them. Section 7.4 presents our results, ending with concluding remarks in Section 7.5.

7.2 Faults and damage in lattice models

The question of how errors spread and propagate in cooperative systems has been studied in a variety of fields. Given the difficulty of creating analytical models for but the simplest systems, most investigations have been conducted by computer simulation, especially in the area of statistical physics of many-body systems. One system that has received much attention is Kauffman's model, which consists of a non-uniform CA with irregular connectivity, in which each cell follows a transition rule that is a random boolean function of the states of its neighbors. The rules, as well as the connections between cells, are randomly selected at the outset, and then remain fixed throughout the system's run (Kauffman, 1993; see also Section 3.4.5). The system has been observed to converge toward limit cycles, after which it can be perturbed by "mutations," which are random changes of rules. Stauffer (1991) and other researchers have studied the spreading of damage in various kinds of two-dimensional lattices (grids) as a function of the probability p of mutating rules within the grid. Critical values of p have been found at which a phase transition seems to occur: above the critical p the damage spreads to the entire lattice, while below it the system is stable with respect to damage spreading.

Another well-known system in which the time evolution of damage has been investigated is the Ising ferromagnet and related spin systems. In these "thermal systems" transition probabilities are a function of the temperature. Stanley et al. (1987) employed Monte Carlo simulations using Metropolis dynamics, finding that there exists a critical temperature T_c, above which (i.e., at high temperatures) an initial damage at a few cells spreads rapidly to the entire system, while below T_c the damage eventually dissipates. Some apparent inconsistencies in this work, due to the use of different transition probability functions, have been resolved by Coniglio et al. (1989).

The general objective of the kind of research summarized above is the study of the temporal limit behavior of the system as a function of some parameter, such as the probability of fault or the thermal noise. For some systems critical behavior has been shown to occur and in some cases critical exponents were computationally determined. For a review of damage dynamics in collective systems from the point of view of computational physics see Jan and de Arcangelis (1994).

7.3   Probabilistic faults in cellular automata

Although the simulation methodology is similar, the main difference between the studies described in the previous section and the work presented herein stems from the fact that we focus on CAs that perform a specified *computational task*, rather than on the long-term dynamics of a physical system under given constraints. From our computational point of view, what is important is the way in which the *task performance* is affected when the system is perturbed.

Our focus is on non-uniform CAs evolved via cellular programming to solve the density and synchronization tasks (see Chapter 4).[1] These CAs advance in time according to prescribed (evolved) *deterministic* rules, however, noise can be introduced, thereby rendering the CAs non-deterministic. For example, for a two-state CA, at each time step the value that is the output of a given deterministic rule can be reversed with probability p_f, denoted the *fault probability*, each cell being treated independently of the others (Figure 7.1). Thus, a cell updates its state in a non-deterministic manner, setting it at the next time step to that specified in the rule table, with probability $1 - p_f$, or the complementary state, with probability p_f. This definition of noise will be used in what follows since it reasonably models the functioning of a multi-component machine in which the computing elements are subject to stochastic transient faults. Other kinds of perturbations are possible, such as cells becoming unavailable ("permanent damage") or switching to another rule for a long, possibly indefinite, period of time. It is also possible to consider the flipping of cell states, either single cells or clusters of cells. Moreover, each cell may be updated at each time step according to one rule with probability p and according to a second rule with probability $1 - p$ (Vichniac et al., 1986). The perturbed Kauffman automata (Stauffer, 1991), in which a cell selects its rule probabilistically, to be then subject to random mutations, is an example similar to ours.

The simulation methodology is based on the concept of "system replicas" (Jan and de Arcangelis, 1994; Wolfram, 1983; Kauffman, 1969). Two systems run in parallel, the original unperturbed one ($p_f = 0$), and a second system, subject to a non-zero probability of error ($p_f > 0$). Both systems start with the same initial configuration at time $t = 0$, after which their temporal behavior is monitored and the Hamming distance between the two configurations at each time step is recorded.[2] This provides us with insight into the faulty CA's behavior, by measuring the amount by which it diverges from a "perfect" computation. Our studies are stochastic in nature, involving a number of measures which are obtained experimentally, averaged over a large number of initial configurations.

---

[1]The evolved CAs discussed in this chapter are fully specified in Appendix C.

[2]The Hamming distance between two configurations is the number of bits by which they differ.

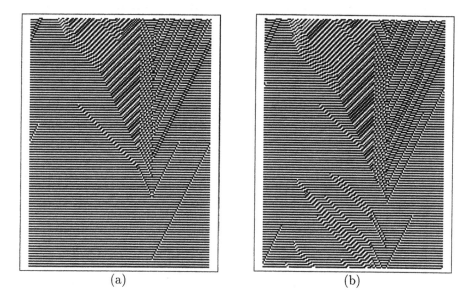

Figure 7.1. The synchronization task: Operation of a coevolved, non-uniform, $r = 1$ CA, with probability of fault $p_f > 0$. (The non-faulty version of this CA is depicted in Figure 4.13b). Grid size is $N = 149$. The pattern of configurations is shown for the first 200 time steps. The initial configurations were generated by randomly setting the state of each grid cell to 0 or 1 with uniform probability. (a) $p_f = 0.0001$. (b) $p_f = 0.001$.

The non-uniform CAs studied are ones that have evolved via cellular programming to perform either the density or synchronization tasks, with our fault-tolerance investigation picking up upon *termination* of the evolutionary process. Figure 7.1 depicts the operation of an evolved CA for two different non-zero p_f values.

7.4 Results

Figure 7.2 depicts the average Hamming distance as a function of the fault probability p_f. We note that the curve is sigmoid-shaped, exhibiting three observable regions: a slow-rising slope ($p_f \leq 0.0005$), followed by a sharp one ($0.0005 < p_f \leq 0.01$), that eventually levels off ($p_f > 0.01$). This latter region exhibits an extremely large Hamming distance, signifying an unacceptable level of computational error. The most important result concerns the first (left-hand) region, which can be considered the *fault-tolerant zone*, where the faulty CA operates in a near-perfect manner. This demonstrates that our evolved CAs exhibit graceful degradation in the face of errors. We also note that there is no essential difference between the two tasks, density and synchronization, except for the higher error level in the "unacceptable" zone, attained by the density CAs. These simulations (as well as the others reported below) were repeated several times, obtaining virtually identical results.

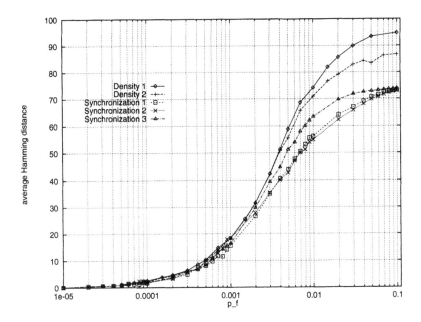

Figure 7.2. Average Hamming distance versus fault probability p_f. Five CAs were studied: two that were evolved to perform the density task, and three that were evolved to perform the synchronization task. Grid size is $N = 149$. For each p_f value the CA under test was run on 1000 randomly generated initial configurations for 300 time steps per configuration. At each time step the Hamming distance between the "perfect" CA and the faulty one is recorded. The average over all configurations and all time steps is represented as a point in the graph.

The above measure furnishes us with our first glimpse into the workings of the faulty CAs, demonstrating an important global characteristic, namely, their ability to tolerate a certain level of faults. We now wish to "zoom" into the fault-tolerant zone, where "good" computational behavior is exhibited, introducing measures to fine-tune our understanding of the faulty CAs' operation. In what follows we shall concentrate on one task, synchronization, due to the improved evolved performance results in comparison to the density task, obtained for the deterministic versions of the CAs (see Chapter 4).[3] We now wish to study the propagation of errors in time, toward which end we examine the Hamming

---

[3]Note that applying the performance measures of Chapter 4 to the *deterministic* versions of the three evolved synchronization CAs discussed herein revealed no differences between them.

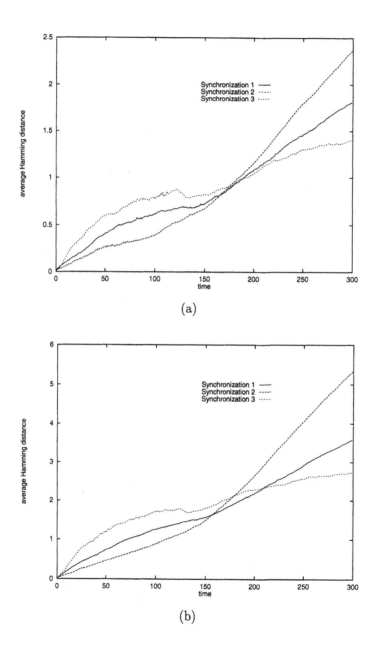

(a)

(b)

Figure 7.3. Hamming distance as a function of time for three CAs that were evolved to perform the synchronization task. Grid size is $N = 149$. Each CA is run on 1000 random initial configurations for 300 time steps per configuration. The Hamming distance per time step is averaged over these configurations. (a) $p_f = 0.00005$. (b) $p_f = 0.0001$.

distance between the perfect and faulty versions, as a function of time (step). Our results are depicted in Figure 7.3. We note that while Hamming distance is limited within the region suggested by Figure 7.2, there are differences between the CAs. Most notable is the high error rate attained by CA 2 in the last 100 time steps.

Further investigation revealed that this is due to *critical zones*. These are specific rules or rule blocks (i.e., blocks of cells containing the same rule) that cause an "avalanche" of error spreading, which may eventually encompass the entire system, as demonstrated in Figure 7.4. Figure 7.4a shows that the CA's error rate peaks around cell 60, which is at the border of rule blocks (see Appendix C). Indeed, when this cell is perturbed (Figure 7.4b), the error may eventually spread to the entire system, resulting in the diminished performance in later time steps, evident in Figure 7.3. Interestingly, this CA has the lowest error rate for the initial part of the computation (Figure 7.3). CA 3 exhibits the opposite time behavior, starting with a higher error rate, which increases, however, more slowly (Figure 7.3). Figure 7.5a shows that this CA exhibits an error peak at the proximity of cell 90, however, a much sharper one than that of CA 2, which partially explains the resulting error containment. Again, cell 90 is at the border of two rule blocks (see Appendix C). Figure 7.5a exhibits a minimum at cell 16, which is also a border cell (between rule blocks), demonstrating that such border rules may act in the opposite manner, "stifling" error spreading rather than enhancing it. CA 1 consists of two major rule blocks, exhibiting different error dispersion behavior, as demonstrated in Figure 7.6. Thus, by introducing time and space measures, we have shown that although all three CAs are within the fault-tolerant zone, their behavior is quite different.

The final issue we consider is that of *recuperation time*. Since our CAs are in effect computational systems, we wish to learn not only whether they recover from faults but also how long this takes. Toward this end we introduced the following measure: the CA of size $N = 149$ is run for 600 time steps with a given fault probability p_f. If the Hamming distance between the perfect and faulty versions passes a certain threshold, which we have set to $0.05N$ bits, at time t_1, and then falls below this threshold at time t_2, staying below for at least three time steps, then recuperation time is defined as $t_2 - t_1$. Note that such "windows" of faulty behavior may occur more than once during the CA's run (i.e., during the 600 time steps); also note that t_2 may equal 600 if the CA never recovers. Simply put, this measure indicates the proportional amount of time that the CA is within a window of unacceptable error level. Our results are depicted in Figure 7.7. For $p_f < 0.0001$ recuperation time is quite short for all three CAs, however, above this fault level, CA 3 exhibits notably higher recuperation time than the other two. It is interesting in that this CA has the lowest error level over time (Figure 7.3).[4] Thus, it is more robust to errors in general, however, certain faults may entail longer recuperation time. This result, along with the

---

[4]Though Figure 7.3 shows results for $p_f \leq 0.0001$, we have verified that the same qualitative behavior is exhibited for $p_f > 0.0001$.

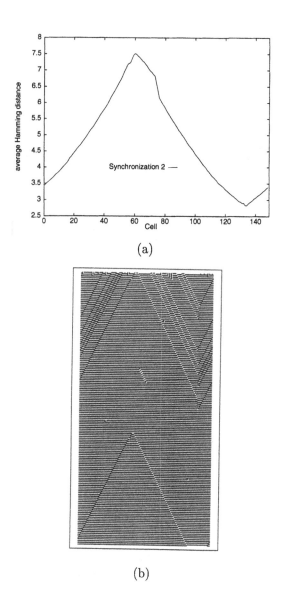

Figure 7.4. Synchronization CA 2: Critical zones. (a) Hamming distance per cell (averaged over 1000 random initial configurations, each run for 300 time steps). Note the peak around cell 60 (the leftmost cell is numbered 0). (b) Perturbing this cell causes an "avalanche" of error spreading. The figure depicts the operation of the CA upon presentation of a random initial configuration. After approximately 200 time steps, cell 60's state is flipped. This cell is situated at the border of rule blocks (see Appendix C). $p_f = 0.0001$ for both (a) and (b). Grid size is $N = 149$.

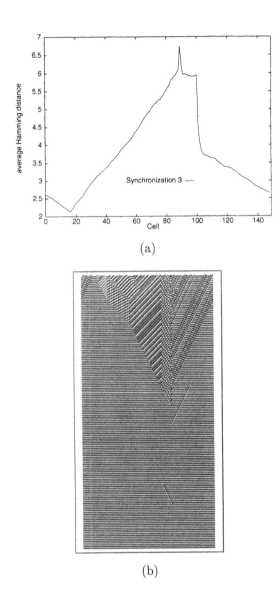

(a)

(b)

Figure 7.5. Synchronization CA 3. (a) Hamming distance per cell (averaged over 1000 random initial configurations, each run for 300 time steps). Note the peak around cell 90, much sharper than that of Figure 7.4. (b) Perturbing this cell does not cause an "avalanche" and the error remains contained. This results in a lower Hamming distance as function of time (Figure 7.3). The figure depicts the operation of the CA upon presentation of a random initial configuration. After approximately 200 time steps, cell 90's state is flipped. This cell is situated at the border of rule blocks (see Appendix C). $p_f = 0.0001$ for both (a) and (b). Grid size is $N = 149$.

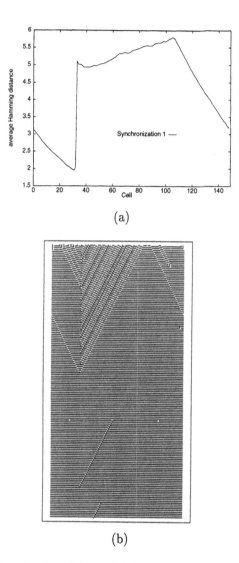

(a)

(b)

Figure 7.6. Synchronization CA 1. (a) Hamming distance per cell (averaged over 1000 random initial configurations, each run for 300 time steps). Two major rule blocks are present, each exhibiting a different error dispersion behavior, the highest error level being that of the "middle" block (note that the left and right blocks contain the same rule, as can be seen in Appendix C, and therefore constitute one block due to the grid's circularity). (b) Three cells are perturbed, in different parts of the grid (cells 20, 70, 120). The error introduced in the middle block propagates, whereas the other two are immediately stifled. The figure depicts the operation of the CA upon presentation of a random initial configuration. After approximately 200 time steps, the states of the above three cells are flipped. $p_f = 0.0001$ for both (a) and (b). Grid size is $N = 149$.

others obtained above, demonstrates the intricate interplay between temporal and spatial factors in our evolved non-uniform CAs.

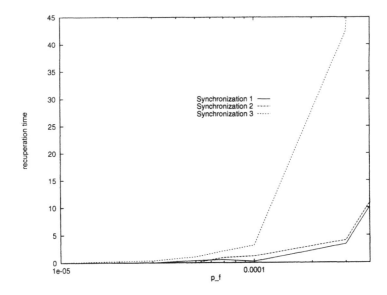

Figure 7.7. Recuperation time as a function of fault probability p_f. Each of the three evolved CAs was run on 1000 random initial configurations for 600 time steps. Average results are depicted in the graph. Grid size is $N = 149$.

7.5   Concluding remarks

We studied the effects of random faults on the behavior of one-dimensional, non-uniform CAs, evolved via cellular programming to perform given computational tasks. Our aim in this chapter was to shed some light on the behavior of such systems under faulty conditions. Using the "system replicas" methodology, involving a comparison between a perfect, non-perturbed version of the CA and a faulty one, we found that our evolved systems exhibit graceful degradation in performance, able to tolerate a certain level of faults. We then zoomed into the fault-tolerant zone, where "good" computational behavior is exhibited, introducing measures to fine-tune our understanding of the faulty CAs' operation. We studied the error level as a function of time and space, as well as the recuperation time needed to recover from faults.

   Our study of evolved non-uniform CAs performing computational tasks revealed an intricate interplay between temporal and spatial factors, with the presence of different rules in the grid giving rise to complex dynamics. Clearly, we have only taken the first step, and there is much yet to be explored. Other types of measures can be considered, such as fault behavior as a function of grid size, permanent faults along with their effects with respect to the rules distribution within the grid, and "total damage time," i.e., the time required for all cells to be damaged at least once. Another interesting issue involves the introduction of faults during the evolutionary process itself to see how well evolution copes with such non-deterministic CAs. Future computing systems may contain thousands or even millions of computing elements. For such large numbers of components, the issue of resilience can no longer be ignored since faults will be likely to occur with high probability. Studies such as the one carried out in this chapter may help deepen our understanding of this important issue.

Chapter 8

Coevolving Architectures for Cellular Machines

Every man is the architect of his own fortune.
Francis Bacon

8.1 Introduction

In the previous chapter we examined the issue of fault tolerance by considering a generalization of the original CA model, involving non-deterministic updating of cell states. In this chapter we generalize on a different aspect of the original model, namely, its standard, homogeneous connectivity. Our investigation is carried out by focusing on the density task (Chapter 4). As a reminder, in this (global) task, the 2-state CA must decide whether or not the initial configuration contains more than 50% 1s, relaxing to a fixed-point pattern of all 1s if the initial density of 1s exceeds 0.5, and all 0s otherwise (e.g., Figure 4.1). Employing the cellular programming algorithm, we found that high-performance systems can be coevolved.

The task was originally studied using locally-connected, one-dimensional grids (Mitchell et al., 1994b; Sipper, 1996a; see Chapter 4). It can be extended in a straightforward manner to two-dimensional, 5-neighbor grids, which posses the same number of local connections per cell as in the one-dimensional, $r = 2$ case. In Section 4.6, having applied our evolutionary algorithm, we found that markedly higher performance is attained for the density task with two-dimensional grids, along with shorter computation times. This finding is intuitively understood by observing that a two-dimensional, locally-connected grid can be embedded in a one-dimensional grid with local and distant connections. This can be achieved, for example, by aligning the rows of the two-dimensional grid so as to form a one-dimensional array; the resulting embedded one-dimensional grid has distant connections of order $\sqrt{N}$, where N is the grid size. Since the density task is global it is likely that the observed superior performance of two-dimensional grids arises from the existence of distant connections that enhance information propagation across the grid.

Motivated by this observation concerning the effect of connection lengths on performance, our primary goal in this chapter is to quantitatively study the relationship between performance and connectivity on a global task, in one-dimensional CAs. The main contribution of this chapter is in identifying the average cellular distance, acd (see next Section), as the prime architectural parameter, which linearly determines CA performance. We find that high-performance architectures can be coevolved concomitantly with the rules, and that it is possible to evolve such architectures that exhibit low connectivity cost as well as high performance (Sipper and Ruppin, 1996a; Sipper and Ruppin, 1996b). Our motivation stems from two primary sources: (1) finding more efficient CA architectures via evolution, and (2) the coevolution of architectures offers a promising approach for solving a general wiring problem for a set of distributed processors, subject to given constraints. The efficient solution of the density task by CAs with evolving architectures may have important applications to designing efficient distributed computing networks.

In the next section we describe the CA architectures studied in this chapter. In Section 8.3 we study CA rule evolution with fixed architectures. In Section 8.4 we extend the cellular programming algorithm, presented in Section 4.3, such that the architecture evolves along with the cellular rules, and in Section 8.5 we study the evolution of low-cost architectures. Our findings, and their possible future application to designing distributed computer networks, are discussed in Section 8.6.

8.2   Architecture considerations

We use the term *architecture* to denote the connectivity pattern of CA cells. As a reminder, in the standard one-dimensional model a cell is connected to r local neighbors on either side, as well as to itself, where r is referred to as the radius (thus, each cell has $2r + 1$ neighbors; see Section 1.2.1). The model we consider is that of non-uniform CAs with *non-standard* architectures, in which cells need not necessarily contain the same rule or be locally connected; however, as with the standard CA model, each cell has a small, identical number of impinging connections. In what follows the term *neighbor* refers to a directly connected cell. We shall employ the cellular programming algorithm to evolve cellular rules for non-uniform CAs, whose architectures are fixed (yet non-standard) during the evolutionary run, or evolve concomitantly with the rules; these are referred to as *fixed* or *evolving* architectures, respectively.

We consider one-dimensional, symmetrical architectures, where each cell has four neighbors, with connection lengths of a and b, as well as a self-connection. Spatially periodic boundary conditions are used, resulting in a circular grid (Figure 8.1). This type of architecture belongs to the general class of *circulant graphs* (Buckley and Harary, 1990): for a given positive integer N, let $n_1, n_2, \ldots, n_k$ be a sequence of integers where

$$0 < n_1 < n_2 < \cdots < n_k < (N+1)/2.$$

Then the *circulant graph* $C_N(n_1, n_2, \ldots, n_k)$ is the graph on N nodes $v_1, v_2, \ldots, v_N$, with node v_i connected to each node $v_{i \pm n_j}$ (mod N). The values n_j are referred to as *connection lengths*. The *distance* between two cells on the circulant is the number of connections one must traverse on the shortest path connecting them. The architectures studied here are circulants $C_N(a, b)$ (Figure 8.1).

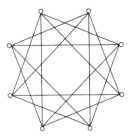

Figure 8.1. A $C_8(2,3)$ circulant graph. Each node is connected to four neighbors, with connection lengths of 2 and 3.

We surmise that attaining high performance on global tasks requires rapid information propagation throughout the CA, and that the rate of information propagation across the grid inversely depends on the average cellular distance (*acd*). Before proceeding to study performance, let us examine how the *acd* of a $C_N(a, b)$ architecture varies as a function of (a, b). As shown in Figure 8.2, the *acd* landscape is extremely rugged (the algorithm used to calculate the *acd* is described in Appendix E). This is due to the relationship between a and b - if $\gcd(a, b) \neq 1$ the *acd* is markedly higher than when $\gcd(a, b) = 1$ (note that the circulant graph $C_N(n_1, n_2, \ldots, n_k)$ is connected if and only if $\gcd(n_1, n_2, \ldots, n_k, N) = 1$, Boesch and Tindell, 1984).

It is straightforward to show that every $C_N(a, b)$ architecture is isomorphic to a $C_N(1, d')$ architecture, for some d', referred to as the *equivalent* d' (see Appendix E). Graph $C_N(a, b)$ is isomorphic to a graph $C_N(1, d')$ if and only if every pair of nodes linked via a connection of length a in $C_N(a, b)$ is linked via a connection of length 1 in $C_N(1, d')$, and every pair linked via a connection of length b in $C_N(a, b)$ is linked via a connection of length d' in $C_N(1, d')$.[1] We may therefore study the performance of $C_N(1, d)$ architectures, our conclusions being applicable to the general $C_N(a, b)$ case. This is important from a practical standpoint since the $C_N(a, b)$ architecture space is extremely large. However, if one wishes to minimize connectivity *cost*, defined as $a + b$, as well as to maximize

---

[1]This is not necessarily a one-to-one mapping. $C_N(a, b)$ may map to $C_N(1, d'_1)$ and $C_N(1, d'_2)$, however, we select the minimum of d'_1 and d'_2, thus obtaining a unique mapping.

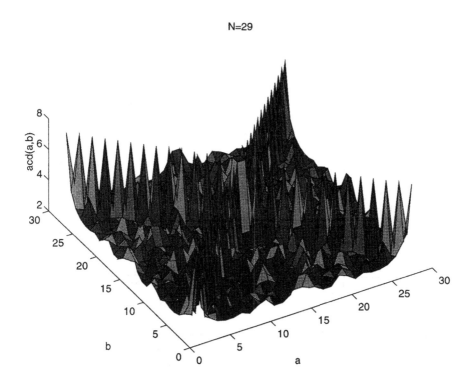

Figure 8.2. The ruggedness of the acd landscape is illustrated by plotting acd as a function of connection lengths (a, b) for grids of size $N = 29$. Each (a, b) pair entails a different $C_{29}(a, b)$ architecture whose acd is represented as a point in the graph.

performance, general $C_N(a, b)$ architectures must be considered (see Section 8.5). The equivalent d' value of a $C_N(a, b)$ architecture may be large, resulting in a lower cost of $C_N(a, b)$ as compared with the isomorphic $C_N(1, d')$ architecture (for example, the equivalent of $C_{101}(3, 5)$ is $C_{101}(1, 32)$).

Figure 8.3 depicts the acd for $C_N(1, d)$ architectures, $N = 101$. It is evident that the acd varies considerably as a function of d; as d increases from $d = 1$, the acd declines and reaches a minimum at $d = O(\sqrt{N})$. This supports the notion put forward in Section 8.1 concerning the advantage of two-dimensional grids.

We concentrate on the following issues:

1. How strongly does the acd determine performance on global tasks? (Section 8.3)

2. Can high-performance architectures be evolved, that is, can "good" d or (a, b) values be discovered through evolution? (Section 8.4)

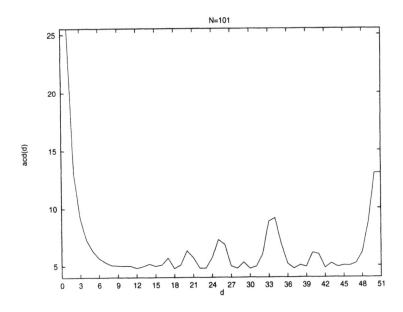

Figure 8.3. $C_{101}(1, d)$: Average cellular distance (acd) as a function of d. acd is plotted for $d \leq N/2$, as it is symmetric about $d = N/2$.

3. Can high-performance architectures be evolved, that exhibit low connectivity cost as well? (Section 8.5)

8.3   Fixed architectures

In this section we study the effects of different architectures on performance, by applying the cellular programming algorithm to the evolution of cellular rules, using fixed, non-standard architectures. We performed numerous evolutionary runs using $C_N(1, d)$ architectures with different values of d, recording the maximal performance attained during the run. As in Chapter 4, *performance* is defined as the average fitness of all grid cells over the last C configurations, normalized to the range $[0.0, 1.0]$ (see discussion in Section 4.4).

Figure 8.4 depicts the results of our evolutionary runs, along with the acd graph. Markedly higher performance is attained for values of d corresponding to low acd values and vice versa. While performance behaves in a rugged, non-monotonic manner as a function of d, it is linearly correlated with acd (with a correlation coefficient of 0.99, and a negligible p value) as depicted in Figure 8.5.

How does the architecture influence performance when the CA is evolved to solve a local task? To test this we introduced the short-lines task: given an initial configuration consisting of five non-filled intervals of random length between $1 - 7$, the CA must reach a final configuration in which the intervals

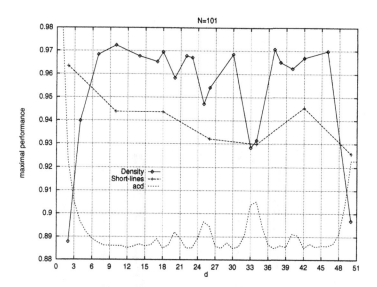

Figure 8.4. $C_{101}(1, d)$: Maximal evolved performance on the density and short-lines tasks as a function of d. The graph represents the average results of 420 evolutionary runs. 21 d values were tested for the density task and 7 for the short-lines task. For each such d value, 15 evolutionary runs were performed with 50,000 initial configurations per run. Each graph point represents the average value of the respective 15 runs. Standard deviations of these averages are in the range $0.003 - 0.011$, i.e., $3\% - 11\%$ of the performance range in question (deviations were computed excluding the two extremal values).

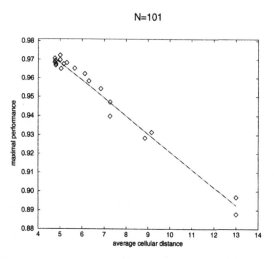

Figure 8.5. $C_{101}(1, d)$: Maximal performance on the density task as a function of average cellular distance. The linear regression shown has a correlation coefficient of 0.99, with a p value that is practically zero.

form continuous lines (Figure 8.6). In this final configuration all cells within the confines of an interval should be in state 1, and all other cells should be in state 0 (in our simulations, cells within an interval in the initial configuration were set to state 1 with probability 0.3; cells outside an interval were set to 0). Figure 8.4 demonstrates that performance for this local task is maximal for minimal d, and decreases as d increases.

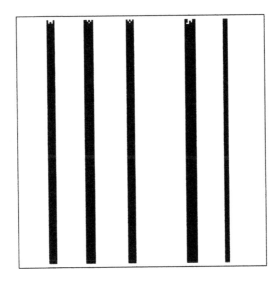

Figure 8.6. The short-lines task: Operation of a coevolved, non-uniform CA of size $N = 149$, with a standard architecture of connectivity radius $r = 2$ ($C_{149}(1, 2)$).

These results demonstrate that performance is strongly dependent upon the architecture, with higher performance attainable by using different architectures than that of the standard CA model. We also observe that the global and local tasks studied have different efficient architectures.

As each $C_N(a, b)$ architecture is isomorphic to a $C_N(1, d)$ one, and since performance is correlated with acd in the $C_N(1, d)$ case, it follows that the performance of general $C_N(a, b)$ architectures is also correlated with acd. It is interesting to note the ruggedness of the *equivalent* d' landscape, depicted in Figure 8.7, representing the equivalent d' value for each (a, b) pair. Table 8.1 presents the performance results of four $C_N(a, b)$ architectures on the density task: $C_{101}(3, 5)$, $C_{102}(3, 5)$, $C_{101}(3, 6)$, and $C_{102}(3, 6)$, demonstrating the dependence on the acd. Since $\gcd(3, 5) = 1$ whereas $\gcd(3, 6) \neq 1$ (resulting in a lower acd for architectures with the former connectivity), we find, as expected, that $C_N(3, 5)$ exhibits significantly higher performance than $C_N(3, 6)$. Furthermore, since $C_{102}(3, 6)$ is not a connected graph (see Section 8.2), this architecture displays even lower performance. The operation of a coevolved, $C_{149}(3, 5)$ CA on the density task is demonstrated in Figure 8.8.

N=29

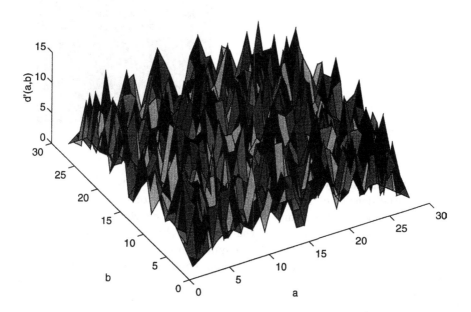

Figure 8.7. The ruggedness of the *equivalent* d' landscape is illustrated by plotting d' as a function of (a, b), for $C_{29}(a, b)$.

(a, b)	N	acd	equivalent d'	mean maximal performance
$(3, 5)$	101	5.98	32	0.96 (0.006)
$(3, 5)$	102	6.02	21	0.96 (0.005)
$(3, 6)$	101	13	2	0.88 (0.01)
$(3, 6)$	102	not connected	none	0.75 (0.07)

Table 8.1. Maximal evolved performance for $C_N(a, b)$ on the density task. For each architecture, 15 evolutionary runs were performed with 50,000 initial configurations per run. The average maximal performance attained on these runs is shown along with standard deviations in parentheses (deviations were computed excluding the two extremal values).

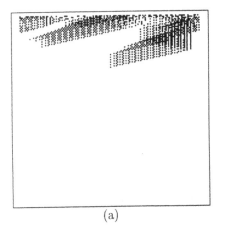

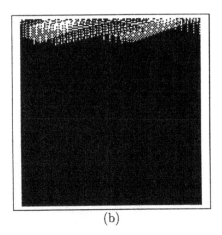

(a)   (b)

Figure 8.8. The density task: Operation of a coevolved, non-uniform, $C_{149}(3,5)$ CA. (a) Initial density of 1s is 0.48. (b) Initial density of 1s is 0.51. Note that computation time, i.e., the number of time steps until convergence to the correct final pattern, is shorter than that of the GKL rule (Figure 4.1). Furthermore, it can be qualitatively observed that the computational "behavior" is different than GKL, as is to be expected due to the different connectivity architecture.

8.4   Evolving architectures

In the previous section we employed the cellular programming algorithm to evolve non-uniform CAs with fixed $C_N(a,b)$ or $C_N(1,d)$ architectures. We concluded that judicious selection of (a,b) or d can notably increase performance, which is highly correlated with the average cellular distance. The question we now pose is whether a-priori specification of the connectivity parameters is indeed necessary or can an efficient architecture coevolve along with the cellular rules. Moreover, can heterogeneous architectures, where each cell may have different d_i or (a_i,b_i) connection lengths, achieve high performance? Below we denote by $C_N(1,d_i)$ and $C_N(a_i,b_i)$ heterogeneous architectures with one or two evolving connection lengths per cell, respectively. Note that these are the cell's input connections, on which information is received; as connectivity is heterogeneous, input and output connections may be different, the latter specified implicitly by the input connections of the neighboring cells.

In order to evolve the architecture along with the rules, the cellular programming algorithm presented in Section 4.3 is modified. Each cell maintains a "genome" consisting of two "chromosomes:" the first, encoding the rule table, is identical to that delineated in Section 4.3, while the second chromosome encodes the cell's connections as Gray-code bit strings (Haykin, 1988).[2] In what

---

[2] A prime characteristic of the Gray code is the adjacency property, i.e., adjacent integers

follows we use grids of size $N = 129$; thus, the architecture chromosome contains 6 bits for evolving $C_{129}(1, d_i)$ architectures and 12 bits for $C_{129}(a_i, b_i)$ architectures. As an example of the latter, if cell i's architecture chromosome equals, say, 000110000100, then it is connected to cells $i \pm 4$ and $i \pm 7$ (mod N), since 000110 and 000100 are the Gray encodings of the decimal values 4 and 7, respectively.

The algorithm now proceeds as in Section 4.3. Initial configurations are presented and fitness scores of each cell are accumulated over C configurations, after which evolution occurs. As with the original algorithm, a cell has access only to its neighbors, and applies genetic operators to the genomes of the fitter ones. Each cell has four connections (in addition to a self-connection), but these need not be identical for all cells, thereby entailing heterogeneous connectivity. We have found that performance can be increased by using slower evolutionary rates for connections than for rules. Thus, while rules evolve every $C = 300$ configurations, connections evolve every $C' = 1500$ configurations. The two-level dynamics engendered by the concomitant evolution of rules and connections markedly increases the size of the space searched by evolution. Our results demonstrate that high performance can be attained, nonetheless.

We performed several evolutionary runs using $C_N(1, d_i)$ architectures, two typical results of which are depicted in Figure 8.9. We find it quite remarkable that the architectures evolved succeed in "selecting" connection lengths d_i that coincide in most cases with minima points of the acd graph, reflecting the strong correlation between performance and acd. This, along with the high levels of performance attained, demonstrates that evolution has succeeded in finding non-uniform CAs with efficient architectures, as well as rules. In fact, the performance attained is higher than that of the fixed-architecture CAs of Section 8.3. Figure 8.10 demonstrates the operation of a coevolved, $C_{129}(1, d_i)$ CA on the density task.

As noted in Section 4.4, Mitchell et al. (1993; 1994b) discussed two possible choices of initial configurations, either uniformly distributed over densities in the range $[0.0, 1.0]$, or binomially distributed over initial densities. As explained therein, this distinction did not prove essential in our studies and we concentrated on the former distribution. Nonetheless, we find that our evolved CAs attain high performance even when applying the more difficult binomial distribution. Observing the results presented in Table 8.2, we note that performance exceeds that of previously evolved CAs, coupled with markedly shorter computation times (as demonstrated, e.g., by Figure 8.10). It is important to note that this is achieved using only 5 connections per cell, as compared to 7, used by the fixed, standard-architecture CAs. It is most likely that our CAs could attain even better results using a higher number of connections per cell, since this entails a notable reduction in acd.

---

differ by a single bit. This is desirable where genetic operators are concerned (Goldberg, 1989).

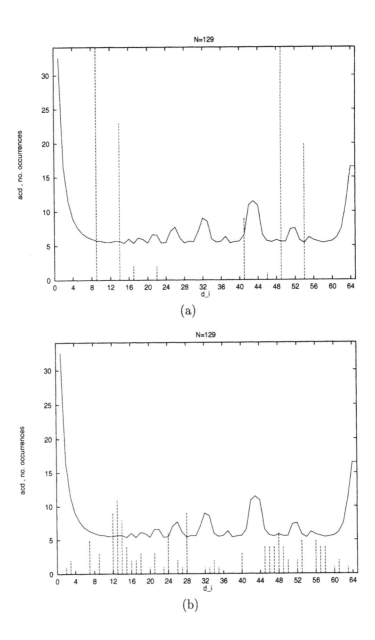

Figure 8.9. Evolving architectures. Results of two typical evolutionary runs using $C_{129}(1, d_i)$. Each figure depicts a histogram of the number of occurrences of evolved d_i values across the grid, overlaid on the acd graph. Performance in both cases is 0.98. Mean d_i value is 31.5 for run (a), 30.8 for run (b).

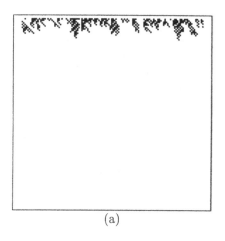
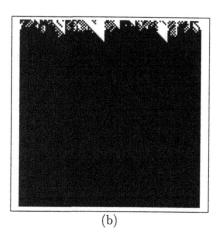

(a)   (b)

Figure 8.10. Density task: Operation of a coevolved, non-uniform, $C_{129}(1, d_i)$ CA. (a) Initial density of 1s is 0.496. (b) Initial density of 1s is 0.504. Note that computation time is shorter than that of the fixed-architecture CA (Figure 8.8), and markedly shorter than that of the GKL rule (Figure 4.1).

desig-nation	rule(s)	architecture	$\#_c$	$\mathcal{P}_{129,10^4}$	$\mathcal{T}_{129,10^4}$
CA (1)	evolved, non-uniform	evolved, non-std.	5	0.791	17
CA (2)	evolved, non-uniform	evolved, non-std.	5	0.788	27
CA (3)	evolved, non-uniform	evolved, non-std.	5	0.781	12
ϕ_{100}	evolved, uniform	fixed, standard	7	0.775	72
ϕ_{11102}	evolved, uniform	fixed, standard	7	0.751	80
ϕ_{17083}	evolved, uniform	fixed, standard	7	0.743	107
GKL	designed, uniform	fixed, standard	7	0.825	74

Table 8.2. A comparison of performance and computation times of the best CAs. $\mathcal{P}_{129,10^4}$ is a measure introduced by Mitchell et al., representing the fraction of correct classifications performed by the CA of size $N = 129$ over 10^4 initial configurations, randomly chosen from a binomial distribution over initial densities. $\mathcal{T}_{129,10^4}$ denotes the average computation time over the 10^4 initial configurations, i.e., the average number of time steps until convergence to the final pattern. $\#_c$ is the number of connections per cell. The CAs designated by (1), (2), and (3), are three of our coevolved CAs; those designated by ϕ_i are CAs reported by Mitchell et al. Coevolved CA (1) is fully specified in Appendix D.

8.5   Evolving low-cost architectures

In the previous section we showed that high-performance architectures can be coevolved using the cellular programming algorithm, thus obviating the need to specify in advance the precise connectivity scheme. The mean d_i value of evolved, $C_{129}(1, d_i)$ architectures was in the range $[30, 40]$ (e.g., Figure 8.9). It is natural to ask whether high-performance architectures can be evolved, which also exhibit low *connectivity cost* per cell, defined as d_i for the $C_N(1, d_i)$ case, and $a_i + b_i$ for $C_N(a_i, b_i)$.

In order to evolve low-cost architectures, we employ the "architectural" cellular programming algorithm of Section 8.4, with a modified cellular fitness value, f_i', incorporating the performance of cell i as well as its connectivity cost:

$$f_i' = f_i - \alpha(a_i + b_i)/N$$

for $C_N(a_i, b_i)$ architectures, and

$$f_i' = f_i - \alpha d_i/N$$

for $C_N(1, d_i)$ ones, where f_i denotes the original fitness value of cell i as defined in Section 4.3, and α is a coefficient in the range $[0.02, 0.04]$. The algorithm now proceeds as in Section 8.4, with an added evolutionary "pressure" toward low-cost architectures.

Figure 8.11 depicts the results of two typical evolutionary runs using $C_N(1, d_i)$ architectures. Comparing this figure with Figure 8.9, we note that low-cost architectures are indeed evolved, exhibiting markedly lower connectivity cost, with only a slight degradation in performance.

In Section 8.2 we observed that every $C_N(a, b)$ architecture is isomorphic to a $C_N(1, d')$ architecture, for some equivalent d'. We noted that general $C_N(a, b)$ architectures come into play when one wishes to minimize connectivity cost, as well as to maximize performance; the equivalent d' value of a $C_N(a, b)$ architecture may be large, resulting in a lower cost of $C_N(a, b)$ as compared with the isomorphic $C_N(1, d')$ architecture. These observations motivated the evolution of general $C_N(a_i, b_i)$ architectures, a typical result of which is demonstrated in Figure 8.12. We note that coevolved, $C_N(a_i, b_i)$ architectures surpass $C_N(1, d_i)$ ones in that better performance is attainable with considerably lower connectivity cost.

8.6   Discussion

In this chapter we studied the relationship between performance and connectivity in evolving, non-uniform CAs. Our main findings are:

1. The performance of fixed-architecture CAs solving global tasks depends strongly and linearly on their average cellular distance. Compared with the standard $C_N(1, 2)$ architecture, considerably higher performance can be

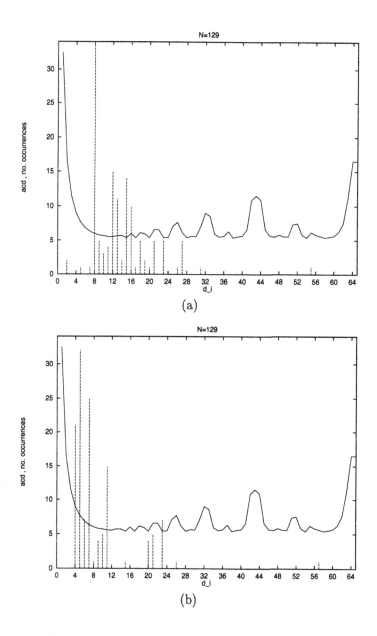

Figure 8.11. Evolving low-cost architectures. Results of two typical evolutionary runs using $C_{129}(1, d_i)$. Each figure depicts a histogram of the number of occurrences of evolved d_i values across the grid, overlaid on the *acd* graph. (a) Performance is 0.97, mean d_i value is 13.6. (b) Performance is 0.96, mean d_i value is 9.

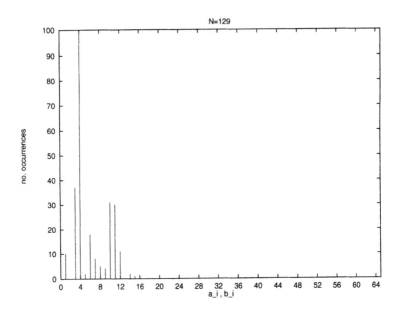

Figure 8.12. Evolving low-cost architectures. Result of a typical evolutionary run using $C_{129}(a_i, b_i)$. The figure depicts a histogram of the number of occurrences of evolved a_i and b_i values across the grid. Performance is 0.97, mean $a_i + b_i$ value is 6.1.

attained at very low connectivity values, by selecting a $C_N(1, d)$ or $C_N(a, b)$ architecture with a low acd value, such that $d, a, b \ll N$.

2. High-performance architectures can be coevolved using the cellular programming algorithm, thus obviating the need to specify in advance the precise connectivity scheme. Furthermore, it is possible to evolve such architectures that exhibit low connectivity cost as well as high performance.

We observed that the average cellular distance landscape is rugged and showed that the performance landscape is qualitatively similar. This suggests an added benefit of evolving, heterogeneous architectures over homogeneous, fixed ones: while the latter may get stuck in a low-performance local minimum, the evolving architectures, where each cell "selects" its own connectivity, result in a melange of local minima, yielding in many cases higher performance.

We have provided empirical evidence as to the added efficiency of $C_N(1, \sqrt{N})$ architectures in solving global tasks, suggesting that the density problem has a good embedding in two dimensions. A theoretical result by Boesch and Wang (1985) states that the minimal diameter of $C_N(a, b)$ circulants is achieved with $C_N(O(\sqrt{N}), O(\sqrt{N}))$. This suggests that the performance landscape has a global maximum at $a, b = O(\sqrt{N})$ (but with $a \neq b$).

We note in passing that as it is physically possible to construct systems of (up to) three dimensions, one can gain the equivalent of long-range connections gratuitously. By this we mean that a physical realization of a locally-connected, three-dimensional system implicitly "contains" a remotely-connected system of lower dimensionality.[3] An interesting extension of our work would be the evolution of architectures using such higher-dimensional grids, which may result in yet better performance, coupled with reduced connectivity cost.

Using our algorithm to solve the density task offers a promising approach for solving a general wiring problem for a set of distributed processors: in this problem one is given a set of processors that should be connected to each other in a way that minimizes average processor distance (i.e., the number of processors a message must traverse on its path between two given processors). Problem constraints may include minimal and maximal connection lengths, prespecified neighbors for some or all cells, and the (possibly distinct) number of impinging connections per processor. Using our algorithm to solve the density task, where each processor is identified with a cell, and connectivity constraints are applied by holding the corresponding connections fixed, will enable the evolution of an efficient wiring scheme for a given distributed computing network, by maximizing the efficiency of global information propagation.

Our simulations have shown that the cellular programming algorithm may degenerate connections. For example, some runs of the short-lines task with evolving $C_N(1, d_i)$ architectures ended up with most cells having $d_i = 0$. This motivates the use of an algorithm that starts out with a large number of connections per cell, to be reduced by evolution, thus yielding increased performance and lower connectivity cost. Ultimately, we wish to attain a system that can adapt to the problem's inherent "landscape."

In summary, this chapter has shed light on the importance of selecting efficient CA architectures, and demonstrated the feasibility of their evolution.

---

[3]As noted, a two-dimensional, locally-connected system cf size N can be embedded in a one-dimensional system with connections of length $\sqrt{N}$. Similarly, a three-dimensional system can be embedded in a two-dimensional system with connections of length $N^{1/3}$, and in a one-dimensional system with connections of length $N^{2/3}$ and $N^{1/3}$.

Chapter 9

Concluding Remarks and Future Research

> *"Would you tell me, please, which way I ought to go from here?"*
> *"That depends a good deal on where you want to get to,"* said the Cat.
>
> Lewis Carroll, *Alice's Adventures in Wonderland*

The parallel cellular machines "designed" by nature exhibit striking problem-solving capabilities. The central question posed in this volume was whether we can mimic nature's achievement, creating artificial machines that exhibit characteristics such as those manifest by their natural counterparts. Clearly, this ambitious goal is yet far off, however, we hope to have taken a small step forward. This chapter ends our tour of parallel cellular machines, in which we studied issues pertaining to their dynamical behavior, the complex computation they exhibit, and the application of artificial evolution to attain such systems.

We selected non-uniform cellular automata as our basic machine model, and showed that universal computation can be attained in simple, non-uniform cellular spaces that are not universal in the uniform case; furthermore, this is accomplished by utilizing a small number of distinct rules (quasi-uniformity). Thus, we demonstrated that simple, non-uniform CAs comprise viable parallel cellular machines.

We then took an artificial-life perspective, presenting a modified non-uniform CA model, with which issues of evolution, adaptation, and multicellularity were studied. We described designed multicellular organisms that display several interesting behaviors, and then turned our attention to evolution in various environments. We concluded that non-uniform CAs and their extensions comprise simple yet versatile models for studying artificial-life phenomena.

In the last and main part of this volume, we asked whether parallel cellular machines can evolve to solve non-trivial, global problems. Toward this end we presented the cellular programming approach, by which such machines locally coevolve to perform computational tasks. We studied in detail a number of problems, some of which suggest possible application domains for our approach, showing that high-performance systems can be evolved. We presented the firefly machine, an evolving, online, autonomous hardware system that implements the cellular programming algorithm, thus demonstrating that "evolving ware," *evolware*, can be attained. The issue of fault tolerance was studied next, looking into the question of robustness, or resilience, namely, how resistant are our evolved machines in the face of errors. We found that they exhibit graceful degradation in performance, able to tolerate a certain level of faults. Finally, we studied non-standard connectivity architectures, showing that these entail definite performance gains, and that, moreover, one can evolve the architecture through a two-level evolutionary process, in which the cellular rules evolve concomitantly with the cellular connections.

The work reported herein represents a first step in an exciting, nascent domain. While results to date are encouraging, there are still several possible avenues of future research, some of which have been explored to a certain extent, while others await to be pursued. Below, we have assembled a list of such future directions (refer also to Section 3.5, where we discussed some additional possible extensions specific to the ALife model presented in Chapter 3):

1. What classes of computational tasks are most suitable for evolving cellular machines? and, what possible applications do they entail? We have noted feasible application areas such as image processing and random number generation. Clearly, more research is necessary in order to elaborate these directions as well as to find new ones.

2. Computation in cellular machines. How are we to understand the emergent, global computation arising in our locally-connected machines? Crutchfield and Mitchell (1995) and Das et al. (1994; 1995) carried out an interesting analysis using automated methods developed by Crutchfield and Young (1989), Hanson and Crutchfield (1992), and Crutchfield and Hanson (1993), for discovering computational structures embedded in the space-time behavior of CAs. Currently, we have performed a more in-depth analysis within the context of our framework in Chapter 4 (see also Sipper, 1996a). This issue is interesting both from a theoretical point of view as well as from a practical one, where it may help guide our search for suitable classes of tasks for such machines.

3. Studying the evolutionary process. The evolutionary algorithms discussed in this volume involve local coevolution, as such presenting novel and interesting dynamics worthy of further study. We wish to enhance our understanding of how evolution creates complex, global behavior in such locally-interconnected systems of simple parts. A first step along this path has been taken herein, for both the ALife model, as well as the cellular programming algorithm.

4. Modifications of the evolutionary algorithms. The representation of CA rules (i.e., the "genome") used in our experiments consists of a bit string, containing a lexicographic listing of all possible neighborhood configurations (see Sections 3.4.1 and 4.2). It has been noted by Land and Belew (1995b) that this representation is fairly low-level and brittle since a change of one bit in the rule table can have a large effect on the computation performed. They evolved uniform CAs to perform the density task using other bit-string representations, as well as a novel, higher-level one, consisting of condition-action pairs; it was demonstrated that better performance is attained when employing the latter. More recently, Andre et al. (1996a; 1996b) used genetic programming (Koza, 1992), in which a rule is represented by a LISP expression, to evolve uniform CAs to perform the density task. This resulted in a CA which outperforms the hand-designed GKL rule (Section 4.2) for certain grid sizes. These experiments demonstrate that changing the bit-string representation, i.e., the encoding of the "genome," may entail notable performance gains; indeed, this issue is of prime import where evolutionary algorithms in general are concerned (for a discussion see, e.g., Mitchell, 1996, Chapter 5). Such encodings could be incorporated into the ALife model of Chapter 3, as well as within the framework of cellular programming. We noted in Section 4.3 that fitness in the cellular programming algorithm is assigned locally to each cell. Another possibility might be to assign fitness scores to blocks of cells, in accordance with their mutual degree of success on the task at hand. Such "block" fitness can also be applied to the ALife model of Chapter 3. It is clear that the novelty of our algorithms leaves much yet to be explored.

5. Modifications of the cellular machine model. In this volume we studied a number of generalizations of the original CA model, including non-uniformity of rules, non-deterministic updating (and its relationship to fault tolerance), non-standard architectures, heterogeneous architectures, and enhanced, "mobile" cellular rules. Other possible avenues to explore include: (1) The application of asynchronous state updating in the cellular programming paradigm, as carried out for the ALife model in Section 3.4.6. A first step along this line was undertaken by Sipper et al. (1996b; 1996c). (2) Three-dimensional grids (and tasks). In this volume we studied one- and two-dimensional grids. Ultimately, three-dimensional systems may be built, enabling new kinds of phenomena to emerge, in analogy to the physical world (de Garis, 1996). As a simple observation consider the fact that signal paths are more collision prone in two dimensions, whereas in three dimensions they may pass each other unperturbed (as an example, consider the mammalian brain). We also noted in Section 8.6 the advantages of three-dimensional systems in terms of signal propagation. Current technology is mostly two-dimensional (e.g., integrated circuits are basically composed of one or more 2D layers), however, future systems, based, e.g.,

on molecular computing (Drexler, 1992), will be three-dimensional.

Our motivation for the above modifications of the cellular machine model partly stems from our desire to attain realistic systems that are more amenable to implementation as evolware.

6. Scaling. As noted in Section 4.7, this involves two separate issues: the evolutionary algorithm and the evolved solutions. We have already explored these to some extent in this volume, though further research is still in order, specifically:

   (a) How does the evolutionary algorithm scale with grid size? Though to date we have performed experiments with different grid sizes, a more in-depth inquiry is needed. Note that as our cellular programming algorithm is local it scales better in terms of hardware resources than the standard (global) genetic algorithm. Adding grid cells requires only local connections in our case, whereas the standard genetic algorithm includes global operators such as fitness ranking and crossover. Indeed, this locality property facilitated the construction of the firefly machine, as noted in Chapter 6. The ALife model of Chapter 3 also exhibits this property.

   (b) How can larger grids be obtained from smaller (evolved) ones, i.e., how can evolved solutions be scaled? This has been purported as an advantage of uniform CAs, since one can directly use the evolved rule in a grid of any desired size. However, this form of *simple* scaling does not bring about *task* scaling; as demonstrated, e.g., by Crutchfield and Mitchell (1995) for the density task, performance decreases as grid size increases. For non-uniform CAs quasi-uniformity may facilitate scaling since only a small number of rules must ultimately be considered. To date, we have attained successful systems using a simple scaling procedure, involving the duplication of the rules grid (Section 5.5), and a more sophisticated scaling approach, which takes into account the evolved grid's local and global structures (Section 4.7).

7. Implementation. As discussed above, this is one of the prime motivations of our work, the goal being to construct evolware.

8. Hierarchy. The idea of decomposing a system into a hierarchy of layers, each carrying out a different function, is ubiquitous in natural as well as artificial systems. As an example of the former, one can cite the human visual system, which begins with low-level image processing in the retina, ending with high-level operations, such as face recognition, performed in the visual cortex. Artificial, feed-forward neural networks are an example of artificial systems exhibiting a layered structure. This idea can be incorporated within our framework, thereby obtaining a hierarchical system,

composed of evolving, layered grids. This could improve the system's performance, facilitate its scaling, and indeed enable entirely new (possibly more difficult) tasks to be confronted.

A related issue is that of the level at which evolution is carried out. For example, our study of architectures consisted of "handing over" to evolution the architectural structure (i.e., the connectivity scheme), in addition to the already present evolving rules structure. In analogy to nature, one can envision the evolution of such structures, that are later "frozen," representing a framework within which evolution must "content" itself. For example, crossover in nature cannot create any conceivable DNA chain, since the (evolved) genomic structure constrains the possible outcomes. A similar point was raised in Section 6.4, upon noting that the firefly machine consists of a fixed part and an evolving part, both specified via configuration strings of the programmable FPGA circuit. We remarked that an interesting outlook on this setup is to consider the evolutionary process as one where an organism evolves within a given species, the former specified by the FPGA's evolving part, the latter specified by the fixed part. This raises the interesting question of whether one can evolve the species itself.

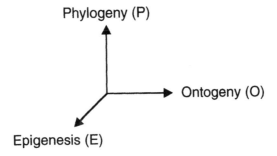

Figure 9.1. The POE model. Partitioning the space of bio-inspired systems along three axes: phylogeny, ontogeny, and epigenesis.

9. If one considers Life on Earth since its very beginning, then the following three levels of organization can be distinguished (Danchin, 1976; Danchin, 1977; Sanchez et al., 1997; Sipper et al., 1997): the *phylogenetic* level concerns the temporal evolution of the genetic programs within individuals and species, the *ontogenetic* level concerns the developmental process of a single multicellular organism, and the *epigenetic* level concerns the learning processes during an individual organism's lifetime, allowing it to integrate the vast quantity of interactions with the outside world (examples of the latter include the nervous system, the immune system, and the endocrine system).

In analogy to nature, the space of bio-inspired systems can be partitioned along these three axes: phylogeny, ontogeny, and epigenesis, giving rise to the *POE* model, schematically depicted in Figure 9.1 (Sanchez et al., 1997; Sipper et al., 1997). As an example, consider the following three paradigms, each positioned along one axis: (P) evolutionary algorithms are the (simplified) artificial counterpart of phylogeny in nature, (O) self-reproducing automata are based on the concept of ontogeny, where a single mother cell gives rise, through multiple divisions, to a multicellular organism, and (E) artificial neural networks embody the epigenetic process, where the system's synaptic weights, and perhaps topological structure, change through interactions with the environment. Within the domains collectively referred to as *soft computing* (Yager and Zadeh, 1994), characterized by ill-defined problems, coupled with the need for continual adaptation or evolution, the above paradigms yield impressive results, often rivaling those of traditional approaches.

The methodologies presented in this volume can be situated along one axis, either the phylogenetic one (e.g., the evolving cellular machines) or the ontogenetic one (e.g., the multicellular systems of Section 3.3). Sanchez et al. (1997) and Sipper et al. (1997) raised the intriguing possibility of creating systems situated within the POE space that exhibit characteristics of two, and ultimately all three axes. This may lead to novel bio-inspired systems, endowed with evolutionary, reproductive, regenerative, and learning capabilities. Thus, enhancing the capacities of the systems described in this volume could result through "infiltration" of the POE space.

Parallel cellular machines hold potential both scientifically, as vehicles for studying phenomena of interest in areas such as complex adaptive systems and artificial life, as well as practically, showing a range of potential future applications, ensuing the construction of adaptive systems. We hope this volume has shed some light on the behavior of such machines, the complex computation they exhibit, and the application of artificial evolution to attain such systems.

Appendix A

Growth and Replication: Specification of Rules

This appendix specifies the rules for the system presented in Section 3.3.4. Note that an A cell dies after attaching a one to the structure, a B cell either dies or spawns a C cell after attachment of zero. All other entries (not shown) of A and B cell rules specify a move to a random vacant cell, while those for C and D cells specify no change.

A cell

    Formation of ones:

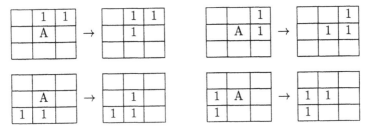

B cell

    Formation of zeros:

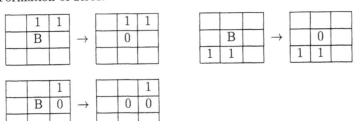

Formation of zero and spawning of C cell:

0	B	
1		

$\rightarrow$

0	0	
1	C	

C cell

Downward movement:

0	0	
1	C	

$\rightarrow$

0	0	
1		
C		

1	1	
1	C	

$\rightarrow$

1	1	
1		
C		

1	1	
0	C	

$\rightarrow$

1	1	
0		
C		

Beginning of upward replication movement and spawning of D cell:

0	0	
	C	

$\rightarrow$

0	0	C
D	0	

Upward replication movement and transfer of one position to the right:

1	1	
0	C	
0		

$\rightarrow$

1	1	C
	0	0

1	1	
1	C	
0	0	

$\rightarrow$

1	1	C
	1	1
0	0	

1	1	
1	C	
1	1	

$\rightarrow$

1	1	C
	1	1
1	1	

0	0	
1	C	
1	1	

$\rightarrow$

0	0	C
	1	1
1	1	

0	0	
1	C	
0	0	

$\rightarrow$

0	0	C
	1	1
0	0	

0	C	
1	1	

$\rightarrow$

		C
	0	0
1	1	

End of upward replication movement:

	C	
0	0	

$\rightarrow$

0	B	

D cell

Move to bottom left-hand side of structure (start position):

0	0			D	0	0
	D	0	→			0

	0			D	0	
	D		→			

0	0			D	0	0
D			→			

Immediate death in case two half structures do not exist:

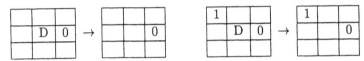

	D	0	→			0

1				1		
	D	0	→			0

Upward replication movement:

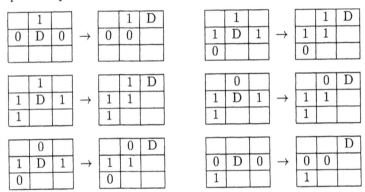

	1				1	D
0	D	0	→	0	0	

	1				1	D
1	D	1	→	1	1	
0				0		

	1				1	D
1	D	1	→	1	1	
1				1		

	0				0	D
1	D	1	→	1	1	
1				1		

	0				0	D
1	D	1	→	1	1	
0				0		

						D
0	D	0	→	0	0	
1				1		

Death after completion of upward movement:

	D					
0			→	0		

Appendix B

A Two-state, r=1 CA that Classifies Density

This appendix is a summary of the result presented by Capcarrere et al. (1996). In the density classification problem, the one-dimensional, two-state CA is presented with an arbitrary initial configuration, and should converge in time to a state of all 1s if the initial configuration contains a density of 1s > 0.5, and to all 0s if this density < 0.5; for an initial density of 0.5, the CA's behavior is undefined. Spatially periodic boundary conditions are used, resulting in a circular grid.

It has been shown that for a uniform one-dimensional grid of fixed size N, and for a fixed radius $r \geq 1$, there exists no two-state CA rule which correctly classifies all possible initial configurations (Land and Belew, 1995a). This says nothing, however, about how well an imperfect CA might perform, one possible approach for obtaining successful CAs being artificial evolution, as described in this volume.

The density classification problem studied to date specifies convergence to one of two fixed-point configurations, which are considered as the output of the computation. Recently, Capcarrere et al. (1996) have shown that a perfect CA density classifier exists, upon defining a different output specification. Consider the uniform, two-state, $r = 1$ rule-184 CA, defined as follows:

$$s_i(t+1) = \begin{cases} s_{i-1}(t) & \text{if } s_i(t) = 0 \\ s_{i+1}(t) & \text{if } s_i(t) = 1 \end{cases}$$

where $s_i(t)$ is the state of cell i at time t. Upon presentation of an arbitrary initial configuration, the grid relaxes to a limit-cycle, within $\lceil N/2 \rceil$ time steps, that provides a classification of the initial configuration's density of 1s: if this density > 0.5 (respectively, < 0.5), then the final configuration consists of one or more blocks of at least two consecutive 1s (0s), interspersed by an alternation of 0s and 1s; for an initial density of exactly 0.5, the final configuration consists of an alternation of 0s and 1s. The computation's output is given by the state of the consecutive block (or blocks) of same-state cells (Figure B.1). As proven by Capcarrere et al.

(1996), this rule performs perfect density classification (including the density=0.5 case). We note in passing that the reflection-symmetric rule 226 holds the same properties of rule 184.

As the input configuration is random, this entails a high Kolmogorov complexity. Intuitively, for a given finite string, this measure concerns the size of the shortest program that computes the string (Li and Vitányi, 1993). Both the fixed-point output of the original problem, as well as the novel "blocks" output, involve a notable reduction with respect to this complexity measure. It has been noted by Mitchell et al. (1994b) that the computational complexity of the input is that of a non-regular language (Hopcroft and Ullman, 1979) since a counter register is needed, whose size is proportional to $\log(N)$, whereas the fixed-point output of the original problem involves a simple regular language (all 0s or all 1s); we note that the novel output specification also involves a regular language (a block of two state-0 or state-1 cells). Capcarrere et al. (1996) thus concluded that their newly proposed density classifier is as viable as the original one with respect to these complexity measures, while surpassing the latter in terms of performance.

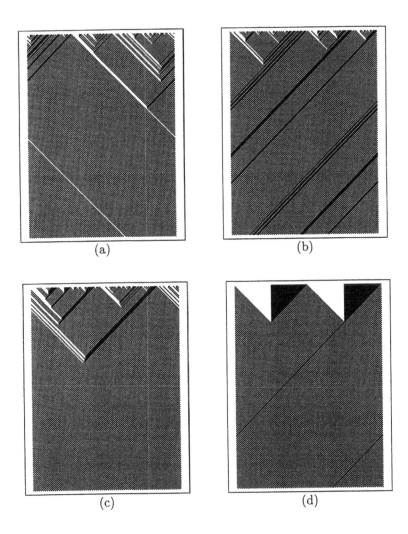

(a) (b)

(c) (d)

Figure B.1. Density classification: Demonstration of the uniform rule-184 CA on four initial configurations. The pattern of configurations is shown for the first 200 time steps. Initial configurations in figures (a)-(c) were randomly generated. (a) Grid size is $N = 149$. Initial density is 0.497, i.e., 75 cells are in state 0, and 74 are in state 1. The final configuration consists of an alternation of 0s and 1s with a single block of two cells in state 0. (b) $N = 149$. Initial density is 0.537. The final configuration consists of an alternation of 0s and 1s with several blocks of two or more cells in state 1. (c) $N = 150$. Initial density is 0.5. The final configuration consists of an alternation of 0s and 1s. (d) $N = 149$. Initial configuration consists of a block of 37 zeros, followed by 37 ones, followed by 37 zeros, ending with 38 ones. The final configuration consists of an alternation of 0s and 1s with a single block of two cells in state 1. In all cases the CA correctly classifies the initial configuration.

Appendix C

Specification of Evolved CAs

The five one-dimensional, non-uniform, $r = 1$ CAs, evolved via cellular programming, and discussed in Chapters 4 and 7, are fully specified in Table C.1. The specification includes the rule found in each cell, where rule numbers are given in accordance with Wolfram's convention, representing the decimal equivalent of the binary number encoding the rule table (as in Figure 4.6). All grid sizes are $N = 149$. Cell 0 is the leftmost cell. Spatially periodic boundary conditions are used, resulting in a circular grid. For an $r = 1$ CA this means that the leftmost and rightmost cells are connected.

Synch. 1:

From cell	To cell	Rule
0	32	31
33	105	83
106	106	19
107	148	31

Synch. 2:

From cell	To cell	Rule
0	55	21
56	56	85
57	58	21
59	60	53
61	73	63
74	132	31
133	148	21

Density 1:

From cell	To cell	Rule
0	39	226
40	40	234
41	71	226
72	72	234
73	142	226
143	144	224
145	148	226

Synch. 3:

From cell	To cell	Rule
0	15	53
16	16	55
17	29	59
30	89	43
90	100	39
101	101	7
102	148	53

Density 2:

From cell	To cell	Rule
0	106	226
107	108	224
109	131	226
132	132	234
133	148	226

Table C.1. Specification of evolved CAs.

Appendix D

Specification of an Evolved Architecture

Coevolved CA (1), whose performance measures are given in Table 8.2, is fully specified in Table D.1. As the architecture in question is non-uniform, $C_{129}(1, d_i)$, this involves 129 rules and d_i values. The 32-bit rule string is shown as 8 hexadecimal digits, with neighborhood configurations given in lexicographic order. The first (leftmost) bit specifies the state to which neighborhood 00000 is mapped to, and so on until the last (rightmost) bit, specifying the state to which neighborhood 11111 is mapped to. The 5 neighborhood bits represent the values of cells $i - d_i, i - 1, i, i + 1, i + d_i \pmod{N}$, respectively. Cell 0 is the leftmost grid cell. Spatially periodic boundary conditions are used, resulting in a circular grid.

Cell	Rule	d_i	Cell	Rule	d_i	Cell	Rule	d_i	Cell	Rule	d_i
0	135107FF	59	33	035117F7	56	66	135107F7	44	99	135107F7	59
1	135107FF	44	34	115107F7	56	67	135107F7	44	100	035117F7	40
2	135107F7	63	35	115107F7	8	68	135107F7	44	101	135117F7	8
3	035107FF	40	36	135107FF	8	69	135107F7	8	102	035117F7	40
4	035107FF	40	37	135107FF	56	70	035107F7	8	103	035107F7	40
5	035107F7	15	38	035107FF	56	71	035117FF	52	104	035107F7	56
6	035117F7	40	39	035107F7	48	72	035107FF	11	105	135107F7	56
7	035107F7	56	40	035107F7	8	73	035107FF	59	106	035105FF	56
8	135117F7	56	41	035107FF	44	74	035107FF	59	107	035117F7	56
9	035107F7	63	42	135107FF	59	75	035107F7	55	108	135117F7	56
10	035107F7	63	43	135107FF	43	76	035117FF	56	109	135117F7	56
11	035107F7	52	44	135107F7	63	77	035107FF	40	110	135107F7	56
12	035127FF	11	45	035107FF	59	78	035107F7	44	111	035107FF	56
13	035127FF	59	46	035117F7	43	79	135117F7	15	112	135107F7	56
14	135117F7	8	47	035107FF	43	80	035107F7	15	113	135107F7	56
15	035107F7	11	48	035107FF	40	81	035107F7	59	114	135107FF	52
16	135117F7	11	49	035117F7	56	82	135107F7	40	115	035107F7	43
17	035117F7	43	50	035105FF	56	83	035107F7	63	116	035107FF	43
18	135107FF	4	51	035107F7	56	84	035107F7	4	117	035107F7	43
19	035117FF	4	52	035107FF	63	85	035127FF	56	118	035107FF	56
20	035117F7	4	53	135107FF	52	86	135107F7	56	119	135107F7	56
21	035117F7	59	54	035105FF	4	87	135107F7	8	120	035107F7	40
22	135107F7	12	55	135107FF	56	88	035157F7	7	121	135107FF	8
23	135107F7	40	56	135107FF	56	89	035117F7	63	122	0510FFF	8
24	135107F7	59	57	035107F7	4	90	035107F7	40	123	035107FF	56
25	035107F7	55	58	035107FF	4	91	035107F7	56	124	135107F7	56
26	135107F7	40	59	135107FF	11	92	035107F7	56	125	035107F7	56
27	035107F7	56	60	135107F7	11	93	035107FF	4	126	035107FF	56
28	035107FF	56	61	035107F7	59	94	035117F7	56	127	035107F7	11
29	035107FF	56	62	035107FF	56	95	135107F7	12	128	135107FF	59
30	035107FF	39	63	135117F7	56	96	035107FF	56			
31	035107F7	56	64	135117F7	48	97	035117FF	63			
32	035117F7	48	65	035117F7	48	98	035107F7	59			

Table D.1. Specification of a CA with an evolved architecture and rules.

Appendix E

Computing *acd* and *equivalent* d'

Determining the diameter and average cellular distance (*acd*) of a general circulant is a difficult problem (Buckley and Harary, 1990). The minimum diameter has been determined for all circulants on N nodes and two connection lengths (Boesch and Wang, 1985). Our interest is in the special case of $C_N(a, b)$. We observe that by symmetry we need only consider the paths from node 0 to each other node j, $j = 1, \ldots, N - 1$ (provided such a path exists). Thus, we express j as $ax + by$ mod N, $x, y \in [-N, N]$ (Boesch and Tindell, 1984). The graphs depicted in Section 8.2 were computed by considering all possible (a, b) pairs. For each such pair, minimum cellular distances from node 0 to all other nodes were computed by considering all possible x, y pairs; the average of these distances was then calculated.

To find the isomorphic $C_N(1, d')$ architecture for a given $C_N(a, b)$ we proceed as follows: consider the list of nodes in the $C_N(a, b)$ graph, $0, 1, \ldots, N - 1$. Now rearrange this list such that nodes originally a units apart are now adjacent (unless $\gcd(a, N) > 1$, in which case b is taken). The equivalent d' is then the minimal number of unit connections to node b from the head of the list (or a, if $\gcd(a, N) > 1$). For example, $C_7(2, 3)$ nodes are rearranged in the following order: $0, 2, 4, 6, 1, 3, 5$, and the equivalent d' value is therefore $d' = 2$ (minimal number of unit connections from node 0 to node 3).

Bibliography

Andre, D., Bennett III, F. H., and Koza, J. R. 1996a. Discovery by genetic programming of a cellular automata rule that is better than any known rule for the majority classification problem. In J. R. Koza, D. E. Goldberg, D. B. Fogel, and R. L. Riolo (eds.), *Genetic Programming 1996: Proceedings of the First Annual Conference*, pp 3–11. The MIT Press, Cambridge, MA.

Andre, D., Bennett III, F. H., and Koza, J. R. 1996b. Evolution of intricate long-distance communication signals in cellular automata using genetic programming. In C. Langton and T. Shimohara (eds.), *Artificial Life V: Proceedings of the Fifth International Workshop on the Synthesis and Simulation of Living Systems*. The MIT Press, Cambridge, MA.

Axelrod, R. 1984. *The Evolution of Cooperation*. Basic Books, Inc., New-York.

Axelrod, R. 1987. The evolution of strategies in the iterated prisoner's dilemma. In L. Davies (ed.), *Genetic Algorithms and Simulated Annealing*, pp 32–42. Pitman, London.

Axelrod, R. and Dion, E. 1988. The further evolution of cooperation. *Science*, vol. 242, pp 1385–1390.

Axelrod, R. and Hamilton, W. D. 1981. The evolution of cooperation. *Science*, vol. 211, pp 1390–1396.

Bäck, T. 1996. *Evolutionary Algorithms in Theory and Practice: Evolution Strategies, Evolutionary Programming, Genetic Algorithms*. Oxford University Press, New York.

Banks, E. R. 1970. Universality in cellular automata. In *IEEE 11th Annual Symposium on Switching and Automata Theory*, pp 194–215. Santa Monica, California.

Bedau, M. A. and Packard, N. H. 1992. Measurement of evolutionary activity, teleology, and life. In C. G. Langton, C. Taylor, J. D. Farmer, and S. Rasmussen (eds.), *Artificial Life II*, vol. X of *SFI Studies in the Sciences of Complexity*, pp 431–461. Addison-Wesley, Redwood City, CA.

Bennett, C. and Grinstein, G. 1985. Role of irreversibility in stabilizing complex and nonenergodic behavior in locally interacting discrete systems. *Physical Review Letters*, vol. 55, pp 657–660.

Berlekamp, E. R., Conway, J. H., and Guy, R. K. 1982. *Winning ways for your mathematical plays*, vol. 2, Chapt. 25, pp 817–850. Academic Press, New York.

Bersini, H. and Detour, V. 1994. Asynchrony induces stability in cellular automata based models. In R. A. Brooks and P. Maes (eds.), *Artificial Life IV*, pp 382–387. The MIT Press, Cambridge, Massachusetts.

Boesch, F. T. and Tindell, R. 1984. Circulants and their connectivities. *Journal of Graph Theory*, vol. 8, pp 487–499.

Boesch, F. T. and Wang, J.-F. 1985. Reliable circulant networks with minimum transmission delay. *IEEE Transactions on Circuits and Systems*, vol. CAS-32, no. 12, pp 1286–1291.

Bonabeau, E. W. and Theraulaz, G. 1994. Why do we need artificial life?. *Artificial Life Journal*, vol. 1, no. 3, pp 303–325. The MIT Press, Cambridge, MA.

Broggi, A., D'Andrea, V., and Destri, G. 1993. Cellular automata as a computational model for low-level vision. *International Journal of Modern Physics C*, vol. 4, no. 1, pp 5–16.

Brooks, R. A. 1991. New approaches to robotics. *Science*, vol. 253, no. 5025, pp 1227–1232.

Buck, J. 1988. Synchronous rhythmic flashing of fireflies II. *The Quarterly Review of Biology*, vol. 63, no. 3, pp 265–289.

Buckley, F. and Harary, F. 1990. *Distance in Graphs*. Addison-Wesley, Redwood City, CA.

Burks, A. (ed.) 1970. *Essays on Cellular Automata*. University of Illinois Press, Urbana, Illinois.

Byl, J. 1989. Self-reproduction in small cellular automata. *Physica D*, vol. 34, pp 295–299.

Cantú-Paz, E. 1995. *A Summary of Research on Parallel Genetic Algorithms*. Technical Report 95007, Illinois Genetic Algorithms Laboratory, University of Illinois at Urbana-Champaign, Urbana, IL.

Capcarrere, M. S., Sipper, M., and Tomassini, M. 1996. A two-state, r=1 cellular automaton that classifies density. *Physical Review Letters*, vol. 77, pp 4969–4971.

Chowdhury, D. R., Gupta, I. S., and Chaudhuri, P. P. 1995. A low-cost high-capacity associative memory design using cellular automata. *IEEE Transactions on Computers*, vol. 44, no. 10, pp 1260–1264.

Codd, E. F. 1968. *Cellular Automata*. Academic Press, New York.

Cohoon, J. P., Hedge, S. U., Martin, W. N., and Richards, D. 1987. Punctuated equilibria: A parallel genetic algorithm. In J. J. Grefenstette (ed.), *Proceedings of the Second International Conference on Genetic Algorithms*, p. 148. Lawrence Erlbaum Associates.

Collins, R. J. and Jefferson, D. R. 1992. AntFarm: Towards simulated evolution. In C. G. Langton, C. Taylor, J. D. Farmer, and S. Rasmussen (eds.), *Artificial Life II*, vol. X of *SFI Studies in the Sciences of Complexity*, pp 579–601. Addison-Wesley, Redwood City, CA.

Coniglio, A., de Arcangelis, L., Herrmann, H. J., and Jan, N. 1989. Exact relations between damage spreading and thermodynamical properties. *Europhysics Letters*, vol. 8, no. 4, pp 315–320.

Connection 1991. *The Connection Machine: CM-200 Series Technical Summary*. Thinking Machines Corporation, Cambridge, Massachusetts.

Coveney, P. and Highfield, R. 1995. *Frontiers of Complexity: The Search for Order in a Chaotic World*. Faber and Faber, London.

Crutchfield, J. P. and Hanson, J. E. 1993. Turbulent pattern bases for cellular automata. *Physica D*, vol. 69, pp 279–301.

Crutchfield, J. P. and Mitchell, M. 1995. The evolution of emergent computation. *Proceedings of the National Academy of Sciences USA*, vol. 92, no. 23, pp 10742–10746.

Crutchfield, J. P. and Young, K. 1989. Inferring statistical complexity. *Physical Review Letters*, vol. 63, pp 105.

Culik II, K., Hurd, L. P., and Yu, S. 1990. Computation theoretic aspects of cellular automata. *Physica D*, vol. 45, pp 357–378.

Danchin, A. 1976. A selective theory for the epigenetic specification of the monospecific antibody production in single cell lines. *Ann. Immunol. (Institut Pasteur)*, vol. 127C, pp 787–804.

Danchin, A. 1977. Stabilisation fonctionnelle et épigénèse: une approche biologique de la genèse de l'identité individuelle. In J.-M. Benoist (ed.), *L'identité*, pp 185–221. Grasset.

Darwin, C. R. 1866. *The Origin of Species*. Penguin, London. 1st edition reprinted, 1968.

Das, R., Crutchfield, J. P., Mitchell, M., and Hanson, J. E. 1995. Evolving globally synchronized cellular automata. In L. J. Eshelman (ed.), *Proceedings of the Sixth International Conference on Genetic Algorithms*, pp 336–343. Morgan Kaufmann, San Francisco, CA.

Das, R., Mitchell, M., and Crutchfield, J. P. 1994. A genetic algorithm discovers particle-based computation in cellular automata. In Y. Davidor, H.-P. Schwefel, and R. Männer (eds.), *Parallel Problem Solving from Nature- PPSN III*, vol. 866 of *Lecture Notes in Computer Science*, pp 344–353. Springer-Verlag, Heidelberg.

Dawkins, R. 1986. *The Blind Watchmaker*. W.W. Norton and Company.

de Garis, H. 1996. "Cam-Brain" ATR's billion neuron artificial brain project: A three year progress report. In *Proceedings of IEEE Third International Conference on Evolutionary Computation (ICEC'96)*, pp 886–891.

Drexler, K. E. 1992. *Nanosystems: Molecular Machinery, Manufacturing and Computation*. John Wiley, New York.

Durand, S., Stauffer, A., and Mange, D. 1994. *Biodule: An Introduction to Digital Biology*. Technical report, Logic Systems Laboratory, Swiss Federal Institute of Technology, Lausanne, Switzerland.

Eldredge, N. and Gould, S. J. 1972. Punctuated equilibria: An alternative to phyletic gradualism. In T. J. M. Schopf (ed.), *Models in Paleobiology*, pp 82–115. Freeman Cooper, San Francisco.

Ermentrout, G. B. and Edelstein-Keshet, L. 1993. Cellular automata approaches to biological modeling. *Journal of Theoretical Biology*, vol. 160, pp 97–133.

Flood, M. M. 1952. *Some experimental games*. Technical Report RM-789-1, The Rand Corporation, Santa Monica, CA.

Fogel, D. B. 1995. *Evolutionary Computation: Toward a New Philosophy of Machine Intelligence*. IEEE Press, Piscataway, NJ.

Fredkin, E. and Toffoli, T. 1982. Conservative logic. *International Journal of Theoretical Physics*, vol. 21, pp 219–253.

Frisch, U., Hasslacher, B., and Pomeau, Y. 1986. Lattice-gas automata for the Navier-Stokes equation. *Physical Review Letters*, vol. 56, pp 1505–1508.

Gacs, P. 1985. Nonergodic one-dimensional media and reliable computation. *Contemporary Mathematics*, vol. 41, pp 125.

Gacs, P., Kurdyumov, G. L., and Levin, L. A. 1978. One-dimensional uniform arrays that wash out finite islands. *Problemy Peredachi Informatsii*, vol. 14, pp 92–98.

Galley, P. and Sanchez, E. 1996. *A hardware implementation of a Tierra processor*. Unpublished internal report (in French), Logic Systems Laboratory, Swiss Federal Institute of Technology, Lausanne.

Gardner, M. 1970. The fantastic combinations of John Conway's new solitaire game "life". *Scientific American*, vol. 223, no. 4, pp 120–123.

Gardner, M. 1971. On cellular automata, self-reproduction, the Garden of Eden and the game "life". *Scientific American*, vol. 224, no. 2, pp 112–117.

Garzon, M. 1990. Cellular automata and discrete neural networks. *Physica D*, vol. 45, pp 431–440.

Goeke, M., Sipper, M., Mange, D., Stauffer, A., Sanchez, E., and Tomassini, M. 1997. Online autonomous evolware. In *Proceedings of The First International Conference on Evolvable Systems: From Biology to Hardware (ICES96)*, Lecture Notes in Computer Science. Springer-Verlag, Heidelberg. (to appear).

Goldberg, D. E. 1989. *Genetic Algorithms in Search, Optimization and Machine Learning*. Addison-Wesley.

Gonzaga de Sá, P. and Maes, C. 1992. The Gacs-Kurdyumov-Levin automaton revisited. *Journal of Statistical Physics*, vol. 67, no. 3/4, pp 507–522.

Gould, S. J. 1982. Darwinism and the expansion of evolutionary theory. *Science*, vol. 216, pp 380–387.

Gould, S. J. and Lewontin, R. C. 1979. The spandrels of San Marco and the Panglossian paradigm: A critique of the adaptationist programme. *Proceedings of the Royal Society of London B*, vol. 205, pp 581–598.

Gould, S. J. and Vrba, E. S. 1982. Exaptation- a missing term in the science of form. *Paleobiology*, vol. 8, pp 4–15.

Guo, Z. and Hall, R. W. 1989. Parallel thinning with two-subiteration algorithms. *Communications of the ACM*, vol. 32, no. 3, pp 359–373.

Gutowitz, H. (ed.) 1990. *Cellular Automata: Theory and Experiment, Proceedings of a Workshop Sponsored by the Center for Nonlinear Studies, Los Alamos National Laboratory, Los Alamos*, vol. 45, Nos. 1-3 of *Physica D*.

Gutowitz, H. and Langton, C. 1995. Mean field theory of the edge of chaos. In F. Morán, A. Moreno, J. J. Merelo, and P. Chacón (eds.), *ECAL'95: Third European Conference on Artificial Life*, vol. 929 of *Lecture Notes in Computer Science*, pp 52–64. Springer-Verlag, Heidelberg.

Hanson, J. E. and Crutchfield, J. P. 1992. The attractor-basin portrait of a cellular automaton. *Journal of Statistical Physics*, vol. 66, pp 1415–1462.

Hardy, J., De Pazzis, O., and Pomeau, Y. 1976. Molecular dynamics of a classical lattice gas: Transport properties and time correlation functions. *Physical Review A*, vol. 13, pp 1949–1960.

Hartman, H. and Vichniac, G. Y. 1986. Inhomogeneous cellular automata. In E. Bienenstock, F. Fogelman, and G. Weisbuch (eds.), *Disordered Systems and Biological Organization*, pp 53–57. Springer-Verlag, Heidelberg.

Haykin, S. 1988. *Digital Communications*. John Wiley and Sons.

Hemmi, H., Mizoguchi, J., and Shimohara, K. 1996. Development and evolution of hardware behaviors. In E. Sanchez and M. Tomassini (eds.), *Towards Evolvable Hardware*, vol. 1062 of *Lecture Notes in Computer Science*, pp 250–265. Springer-Verlag, Heidelberg.

Hernandez, G. and Herrmann, H. J. 1996. Cellular-automata for elementary image-enhancement. *CVGIP: Graphical Models and Image Processing*, vol. 58, no. 1, pp 82–89.

Higuchi, T., Iwata, M., Kajitani, I., Iba, H., Hirao, Y., Furuya, T., and Manderick, B. 1996. Evolvable hardware and its application to pattern recognition and fault-tolerant systems. In E. Sanchez and M. Tomassini (eds.), *Towards Evolvable Hardware*, vol. 1062 of *Lecture Notes in Computer Science*, pp 118–135. Springer-Verlag, Heidelberg.

Holland, J. H. 1975. *Adaptation in Natural and Artificial Systems*. The University of Michigan Press, Ann Arbor, Michigan.

Hopcroft, J. E. and Ullman, J. D. 1979. *Introduction to Automata Theory Languages and Computation*. Addison-Wesley, Redwood City, CA.

Hortensius, P. D., McLeod, R. D., and Card, H. C. 1989a. Parallel random number generation for VLSI systems using cellular automata. *IEEE Transactions on Computers*, vol. 38, no. 10, pp 1466–1473.

Hortensius, P. D., McLeod, R. D., Pries, W., Miller, D. M., and Card, H. C. 1989b. Cellular automata-based pseudorandom number generators for built-in self-test. *IEEE Transactions on Computer-Aided Design*, vol. 8, no. 8, pp 842–859.

Huberman, B. A. and Glance, N. S. 1993. Evolutionary games and computer simulations. *Proceedings of the National Academy of Sciences USA*, vol. 90, pp 7716–7718.

Iwata, M., Kajitani, I., Yamada, H., Iba, H., and Higuchi, T. 1996. A pattern recognition system using evolvable hardware. In H.-M. Voigt, W. Ebeling, I. Rechenberg, and H.-P. Schwefel (eds.), *Parallel Problem Solving from Nature - PPSN IV*, vol. 1141 of *Lecture Notes in Computer Science*, pp 761–770. Springer-Verlag, Heidelberg.

Jan, N. and de Arcangelis, L. 1994. Computational aspects of damage spreading. In D. Stauffer (ed.), *Annual Reviews of Computational Physics*, vol. I, pp 1–16. World Scientific, Singapore.

Jefferson, D., Collins, R., Cooper, C., Dyer, M., Flowers, M., Korf, R., Taylor, C., and Wang, A. 1992. Evolution as a theme in Artificial Life: The Genesys/Tracker system. In C. G. Langton, C. Taylor, J. D. Farmer, and S. Rasmussen (eds.), *Artificial Life II*, vol. X of *SFI Studies in the Sciences of Complexity*, pp 549–578. Addison-Wesley, Redwood City, CA.

Joyce, G. F. 1989. RNA evolution and the origins of life. *Nature*, vol. 338, pp 217–224.

Kaneko, K., Tsuda, I., and Ikegami, T. (eds.) 1994. *Constructive Complexity and Artificial Reality, Proceedings of the Oji International Seminar on Complex Systems- from Complex Dynamical Systems to Sciences of Artificial Reality*, vol. 75, Nos. 1-3 of *Physica D*.

Kauffman, S. A. 1969. Metabolic stability and epigenesis in randomly constructed genetic nets. *Journal of Theoretical Biology*, vol. 22, pp 437–467.

Kauffman, S. A. 1993. *The Origins of Order*. Oxford University Press, New York.

Kauffman, S. A. and Johnsen, S. 1992. Co-evolution to the edge of chaos: Coupled fitness landscapes, poised states, and co-evolutionary avalanches. In C. G. Langton, C. Taylor, J. D. Farmer, and S. Rasmussen (eds.), *Artificial Life II*, vol. X of *SFI Studies in the Sciences of Complexity*, pp 325–369. Addison-Wesley, Redwood City, CA.

Kauffman, S. A. and Weinberger, E. D. 1989. The NK model of rugged fitness landscapes and its application to maturation of the immune response. *Journal of Theoretical Biology*, vol. 141, pp 211–245.

Kitano, H. 1996. Morphogenesis for evolvable systems. In E. Sanchez and M. Tomassini (eds.), *Towards Evolvable Hardware*, vol. 1062 of *Lecture Notes in Computer Science*, pp 99–117. Springer-Verlag, Heidelberg.

Knuth, D. E. 1981. *The Art of Computer Programming: Volume 2, Seminumerical Algorithms*. Addison-Wesley, Reading, MA, second edition.

Kohavi, Z. 1970. *Switching and Finite Automata Theory*. McGraw-Hill Book Company.

Koza, J. R. 1992. *Genetic Programming*. The MIT Press, Cambridge, Massachusetts.

Koza, J. R., Bennett III, F. H., Andre, D., and Keane, M. A. 1996. Automated WYWIWYG design of both the topology and component values of electrical circuits using genetic programming. In J. R. Koza, D. E. Goldberg, D. B. Fogel, and R. L. Riolo (eds.), *Genetic Programming 1996: Proceedings of the First Annual Conference*, pp 123–131. The MIT Press, Cambridge, MA.

Land, M. and Belew, R. K. 1995a. No perfect two-state cellular automata for density classification exists. *Physical Review Letters*, vol. 74, no. 25, pp 5148–5150.

Land, M. and Belew, R. K. 1995b. Towards a self-replicating language for computation. In J. R. McDonnell, R. G. Reynolds, and D. B. Fogel (eds.), *Evolutionary programming IV: Proceedings of the Fourth Annual Conference on Evolutionary Programming*, pp 403–413. The MIT Press, Cambridge, Massachusetts.

Langton, C. G. 1984. Self-reproduction in cellular automata. *Physica D*, vol. 10, pp 135–144.

Langton, C. G. 1986. Studying artificial life with cellular automata. *Physica D*, vol. 22, pp 120–140.

Langton, C. G. (ed.) 1989. *Artificial Life*, vol. VI of *SFI Studies in the Sciences of Complexity*. Addison-Wesley, Redwood City, CA.

Langton, C. G. 1990. Computation at the edge of chaos: Phase transitions and emergent computation. *Physica D*, vol. 42, pp 12–37.

Langton, C. G. 1992a. Life at the edge of chaos. In C. G. Langton, C. Taylor, J. D. Farmer, and S. Rasmussen (eds.), *Artificial Life II*, vol. X of *SFI Studies in the Sciences of Complexity*, pp 41–91. Addison-Wesley, Redwood City, CA.

Langton, C. G. 1992b. Preface. In C. G. Langton, C. Taylor, J. D. Farmer, and S. Rasmussen (eds.), *Artificial Life II*, vol. X of *SFI Studies in the Sciences of Complexity*, pp xiii–xviii. Addison-Wesley, Redwood City, CA.

Langton, C. G. 1994. Editor's introduction. *Artificial Life Journal*, vol. 1, no. 1/2, pp v–viii. The MIT Press, Cambridge, MA.

Langton, C. G., Taylor, C., Farmer, J. D., and Rasmussen, S. (eds.) 1992. *Artificial Life II*, vol. X of *SFI Studies in the Sciences of Complexity*. Addison-Wesley, Redwood City, CA.

Levy, S. 1992. *Artificial Life: The Quest for a New Creation*. Random House.

Li, M. and Vitányi, P. 1993. *An Introduction to Kolmogorov Complexity and its Applications*. Springer-Verlag, New-York.

Li, W., Packard, N. H., and Langton, C. G. 1990. Transition phenomena in cellular automata rule space. *Physica D*, vol. 45, pp 77–94.

Lindgren, K. 1992. Evolutionary phenomena in simple dynamics. In C. G. Langton, C. Taylor, J. D. Farmer, and S. Rasmussen (eds.), *Artificial Life II*, vol. X of *SFI Studies in the Sciences of Complexity*, pp 295–312. Addison-Wesley, Redwood City, CA.

Lindgren, K. and Nordahl, M. G. 1990. Universal computation in simple one-dimensional cellular automata. *Complex Systems*, vol. 4, pp 299–318.

Lindgren, K. and Nordahl, M. G. 1994a. Cooperation and community structure in artificial ecosystems. *Artificial Life Journal*, vol. 1, no. 1/2, pp 15–37. The MIT Press, Cambridge, MA.

Lindgren, K. and Nordahl, M. G. 1994b. Evolutionary dynamics of spatial games. *Physica D*, vol. 75, pp 292–309.

Lumer, E. D. and Nicolis, G. 1994. Synchronous versus asynchronous dynamics in spatially distributed systems. *Physica D*, vol. 71, pp 440–452.

Manderick, B. and Spiessens, P. 1989. Fine-grained parallel genetic algorithms. In J. D. Schaffer (ed.), *Proceedings of the Third International Conference on Genetic Algorithms*, p. 428. Morgan Kaufmann.

Mange, D., Goeke, M., Madon, D., Stauffer, A., Tempesti, G., and Durand, S. 1996. Embryonics: A new family of coarse-grained Field-Programmable Gate Arrays with self-repair and self-reproducing properties. In E. Sanchez and M. Tomassini (eds.), *Towards Evolvable Hardware*, vol. 1062 of *Lecture Notes in Computer Science*, pp 197–220. Springer-Verlag, Heidelberg. Also available as: Technical Report 95/154, Department of Computer Science, Swiss Federal Institute of Technology, Lausanne, Switzerland, November, 1995.

Mange, D., Sanchez, E., Stauffer, A., Tempesti, G., Durand, S., Marchal, P., and Piguet, C. 1995. *Embryonics: A New Methodology for Designing Field-Programmable Gate Arrays with Self-Repair and Self-Reproducing Properties.* Technical Report 95/152, Department of Computer Science, Swiss Federal Institute of Technology, Lausanne, Switzerland.

Mange, D. and Stauffer, A. 1994. Introduction to embryonics: Towards new self-repairing and self-reproducing hardware based on biological-like properties. In N. M. Thalmann and D. Thalmann (eds.), *Artificial Life and Virtual Reality*, pp 61–72. John Wiley, Chichester, England.

Marchal, P., Nussbaum, P., Piguet, C., and Sipper, M. 1997. Speeding up digital ecologies evolution using a hardware emulator: Preliminary results. In *Proceedings of The First International Conference on Evolvable Systems: From Biology to Hardware (ICES96)*, Lecture Notes in Computer Science. Springer-Verlag, Heidelberg. (to appear).

Marchal, P., Piguet, C., Mange, D., Stauffer, A., and Durand, S. 1994. Embryological development on silicon. In R. A. Brooks and P. Maes (eds.), *Artificial Life IV*, pp 365–370. The MIT Press, Cambridge, Massachusetts.

Margolus, N. 1984. Physics-like models of computation. *Physica D*, vol. 10, pp 81–95.

Mayr, E. 1976. *Evolution and the Diversity of Life.* Harvard University Press, Cambridge, MA.

Mayr, E. 1982. *The Growth of Biological Thought.* Harvard University Press, Cambridge, MA.

Michalewicz, Z. 1996. *Genetic Algorithms + Data Structures = Evolution Programs.* Springer-Verlag, Heidelberg, third edition.

Miller, S. L. 1953. A production of amino acids under possible primitive Earth conditions. *Science,* vol. 117, pp 528–529.

Miller, S. L. and Urey, H. C. 1959. Organic compound synthesis on the primitive Earth. *Science,* vol. 130, no. 3370, pp 245–251.

Millman, J. and Grabel, A. 1987. *Microelectronics.* McGraw-Hill Book Company, second edition.

Minsky, M. L. 1967. *Computation: Finite and Infinite Machines.* Prentice-Hall, Englewood Cliffs, New Jersey.

Mitchell, M. 1996. *An Introduction to Genetic Algorithms.* MIT Press, Cambridge, MA.

Mitchell, M., Crutchfield, J. P., and Hraber, P. T. 1994a. Dynamics, computation, and the "edge of chaos": A re-examination. In G. Cowan, D. Pines, and D. Melzner (eds.), *Complexity: Metaphors, Models and Reality,* pp 491–513. Addison-Wesley, Reading, MA.

Mitchell, M., Crutchfield, J. P., and Hraber, P. T. 1994b. Evolving cellular automata to perform computations: Mechanisms and impediments. *Physica D,* vol. 75, pp 361–391.

Mitchell, M., Hraber, P. T., and Crutchfield, J. P. 1993. Revisiting the edge of chaos: Evolving cellular automata to perform computations. *Complex Systems,* vol. 7, pp 89–130.

Mueller, L. D. and Feldman, M. W. 1988. The evolution of altruism by kin selection: New phenomena with strong selection. *Ethology and Sociobiology,* vol. 9, pp 223–240.

Murakawa, M., Yoshizawa, S., Kajitani, I., Furuya, T., Iwata, M., and Higuchi, T. 1996. Hardware evolution at function level. In H.-M. Voigt, W. Ebeling, I. Rechenberg, and H.-P. Schwefel (eds.), *Parallel Problem Solving from Nature - PPSN IV,* vol. 1141 of *Lecture Notes in Computer Science,* pp 62–71. Springer-Verlag, Heidelberg.

Nourai, F. and Kashef, R. S. 1975. A universal four-state cellular computer. *IEEE Transactions on Computers,* vol. c-24, no. 8, pp 766–776.

Nowak, M. A., Bonhoeffer, S., and May, R. M. 1994. Spatial games and the maintenance of cooperation. *Proceedings of the National Academy of Sciences USA*, vol. 91, pp 4877–4881.

Nowak, M. A. and May, R. M. 1992. Evolutionary games and spatial chaos. *Nature*, vol. 359, pp 826–829.

Packard, N. H. 1988. Adaptation toward the edge of chaos. In J. A. S. Kelso, A. J. Mandell, and M. F. Shlesinger (eds.), *Dynamic Patterns in Complex Systems*, pp 293–301. World Scientific, Singapore.

Pagels, H. R. 1989. *The Dreams of Reason: The Computer and the Rise of the Sciences of Complexity*. Bantam Books, New York.

Park, S. K. and Miller, K. W. 1988. Random number generators: Good ones are hard to find. *Communications of the ACM*, vol. 31, no. 10, pp 1192–1201.

Perrier, J.-Y., Sipper, M., and Zahnd, J. 1996. Toward a viable, self-reproducing universal computer. *Physica D*, vol. 97, pp 335–352.

Poundstone, W. 1992. *The Prisoner's Dilemma*. Doubleday, New York.

Preston, Jr., K. and Duff, M. J. B. 1984. *Modern Cellular Automata: Theory and Applications*. Plenum Press, New York.

Pries, W., Thanailakis, A., and Card, H. C. 1986. Group properties of cellular automata and VLSI applications. *IEEE Transactions on Computers*, vol. C-35, no. 12, pp 1013–1024.

Rasmussen, S., Knudsen, C., and Feldberg, R. 1992. Dynamics of programmable matter. In C. G. Langton, C. Taylor, J. D. Farmer, and S. Rasmussen (eds.), *Artificial Life II*, vol. X of *SFI Studies in the Sciences of Complexity*, pp 211–254. Addison-Wesley, Redwood City, CA.

Ray, T. S. 1992. An approach to the synthesis of life. In C. G. Langton, C. Taylor, J. D. Farmer, and S. Rasmussen (eds.), *Artificial Life II*, vol. X of *SFI Studies in the Sciences of Complexity*, pp 371–408. Addison-Wesley, Redwood City, CA.

Ray, T. S. 1994a. An evolutionary approach to synthetic biology: Zen and the art of creating life. *Artificial Life Journal*, vol. 1, no. 1/2, pp 179–209. The MIT Press, Cambridge, MA.

Ray, T. S. 1994b. *A Proposal to Create a Network-Wide Biodiversity Reserve for Digital Organisms*. unpublished. see also: *Science*, vol. 264, May, 1994, page 1085.

Reggia, J. A., Armentrout, S. L., Chou, H.-H., and Peng, Y. 1993. Simple systems that exhibit self-directed replication. *Science*, vol. 259, pp 1282–1287.

Sanchez, E. 1996. Field-Programmable Gate Array (FPGA) circuits. In E.

Sanchez and M. Tomassini (eds.), *Towards Evolvable Hardware*, vol. 1062 of *Lecture Notes in Computer Science*, pp 1–18. Springer-Verlag, Heidelberg.

Sanchez, E., Mange, D., Sipper, M., Tomassini, M., Pérez-Uribe, A., and Stauffer, A. 1997. Phylogeny, ontogeny, and epigenesis: Three sources of biological inspiration for softening hardware. In *Proceedings of The First International Conference on Evolvable Systems: From Biology to Hardware (ICES96)*, Lecture Notes in Computer Science. Springer-Verlag, Heidelberg. (to appear).

Sanchez, E. and Tomassini, M. (eds.) 1996. *Towards Evolvable Hardware*, vol. 1062 of *Lecture Notes in Computer Science*. Springer-Verlag, Heidelberg.

Schwefel, H.-P. 1995. *Evolution and Optimum Seeking*. John Wiley & Sons, New York.

Simon, H. 1969. *The Sciences of the Artificial*. The MIT Press, Cambridge, Massachusetts.

Sipper, M. 1994. Non-uniform cellular automata: Evolution in rule space and formation of complex structures. In R. A. Brooks and P. Maes (eds.), *Artificial Life IV*, pp 394–399. The MIT Press, Cambridge, Massachusetts.

Sipper, M. 1995a. An introduction to artificial life. *Explorations in Artificial Life (special issue of AI Expert)*, pp 4–8. Miller Freeman, San Francisco, CA.

Sipper, M. 1995b. Quasi-uniform computation-universal cellular automata. In F. Morán, A. Moreno, J. J. Merelo, and P. Chacón (eds.), *ECAL'95: Third European Conference on Artificial Life*, vol. 929 of *Lecture Notes in Computer Science*, pp 544–554. Springer-Verlag, Heidelberg.

Sipper, M. 1995c. Studying artificial life using a simple, general cellular model. *Artificial Life Journal*, vol. 2, no. 1, pp 1–35. The MIT Press, Cambridge, MA.

Sipper, M. 1996a. Co-evolving non-uniform cellular automata to perform computations. *Physica D*, vol. 92, pp 193–208.

Sipper, M. 1996b. The evolution of parallel cellular machines: Toward evolware. *BioSystems*. (to appear).

Sipper, M. 1997a. Designing evolware by cellular programming. In *Proceedings of The First International Conference on Evolvable Systems: From Biology to Hardware (ICES96)*, Lecture Notes in Computer Science. Springer-Verlag, Heidelberg. (to appear).

Sipper, M. 1997b. Evolving uniform and non-uniform cellular automata networks. In D. Stauffer (ed.), *Annual Reviews of Computational Physics*, vol. V. World Scientific, Singapore. (to appear).

Sipper, M. and Ruppin, E. 1996a. Co-evolving architectures for cellular machines. *Physica D*. (to appear).

Sipper, M. and Ruppin, E. 1996b. Co-evolving cellular architectures by cellular programming. In *Proceedings of IEEE Third International Conference on Evolutionary Computation (ICEC'96)*, pp 306–311.

Sipper, M., Sanchez, E., Mange, D., Tomassini, M., Pérez-Uribe, A., and Stauffer, A. 1997. Softening hardware via phylogenetic, ontogenetic, and epigenetic computation. (submitted).

Sipper, M. and Tomassini, M. 1996a. Co-evolving parallel random number generators. In H.-M. Voigt, W. Ebeling, I. Rechenberg, and H.-P. Schwefel (eds.), *Parallel Problem Solving from Nature - PPSN IV*, vol. 1141 of *Lecture Notes in Computer Science*, pp 950–959. Springer-Verlag, Heidelberg.

Sipper, M. and Tomassini, M. 1996b. Generating parallel random number generators by cellular programming. *International Journal of Modern Physics C*, vol. 7, no. 2, pp 181–190.

Sipper, M., Tomassini, M., and Beuret, O. 1996a. Studying probabilistic faults in evolved non-uniform cellular automata. *International Journal of Modern Physics C*, vol. 7, no. 6.

Sipper, M., Tomassini, M., and Capcarrere, M. 1996b. Designing cellular automata using a parallel evolutionary algorithm. *Nuclear Instruments & Methods in Physics Research, Section A*. (to appear).

Sipper, M., Tomassini, M., and Capcarrere, M. 1996c. Evolving asynchronous and scalable non-uniform cellular automata. (submitted).

Smith, A. 1969. *Cellular automata theory*. Technical Report 2, Stanford Electronic Lab., Stanford University.

Smith, A. R. 1971. Simple computation-universal cellular spaces. *Journal of ACM*, vol. 18, pp 339–353.

Smith, A. R. 1992. Simple nontrivial self-reproducing machines. In C. G. Langton, C. Taylor, J. D. Farmer, and S. Rasmussen (eds.), *Artificial Life II*, vol. X of *SFI Studies in the Sciences of Complexity*, pp 709–725. Addison-Wesley, Redwood City, CA.

Stanley, H. E., Stauffer, D., Kertész, J., and Herrmann, H. J. 1987. Dynamics of spreading phenomena in two-dimensional Ising models. *Physical Review Letters*, vol. 59, no. 20, pp 2326–2328.

Starkweather, T., Whitley, D., and Mathias, K. 1991. Optimization using distributed genetic algorithms. In H.-P. Schwefel and R. Männer (eds.), *Parallel Problem Solving from Nature*, vol. 496 of *Lecture Notes in Computer Science*, p. 176. Springer-Verlag, Heidelberg.

Stauffer, D. 1991. Computer simulations of cellular automata. *Journal Of Physics A: Mathematical And General*, vol. 24, pp 909–927.

Stauffer, D. and de Arcangelis, L. 1996. Dynamics and strong size effects of a bootstrap percolation problem. *International Journal of Modern Physics C*, vol. 7, pp 739–745.

Steels, L. 1994. The artificial life roots of artificial intelligence. *Artificial Life Journal*, vol. 1, no. 1/2, pp 75–110. The MIT Press, Cambridge, MA.

Stork, D. G., Jackson, B., and Walker, S. 1992. "Non-optimality" via pre-adaptation in simple neural systems. In C. G. Langton, C. Taylor, J. D. Farmer, and S. Rasmussen (eds.), *Artificial Life II*, vol. X of *SFI Studies in the Sciences of Complexity*, pp 409–429. Addison-Wesley, Redwood City, CA.

Strogatz, S. H. and Stewart, I. 1993. Coupled oscillators and biological synchronization. *Scientific American*, vol. 269, no. 6, pp 102–109.

Tanese, R. 1987. Parallel genetic algorithms for a hypercube. In J. J. Grefenstette (ed.), *Proceedings of the Second International Conference on Genetic Algorithms*, p. 177. Lawrence Erlbaum Associates.

Taylor, C. and Jefferson, D. 1994. Artificial life as a tool for biological inquiry. *Artificial Life Journal*, vol. 1, no. 1/2, pp 1–13. The MIT Press, Cambridge, MA.

Tempesti, G. 1995. A new self-reproducing cellular automaton capable of construction and computation. In F. Morán, A. Moreno, J. J. Merelo, and P. Chacón (eds.), *ECAL'95: Third European Conference on Artificial Life*, vol. 929 of *Lecture Notes in Computer Science*, pp 555–563. Springer-Verlag, Heidelberg.

Thompson, A. 1997. An evolved circuit, intrinsic in silicon, entwined with physics. In *Proceedings of The First International Conference on Evolvable Systems: From Biology to Hardware (ICES96)*, Lecture Notes in Computer Science. Springer-Verlag, Heidelberg. (to appear).

Thompson, A., Harvey, I., and Husbands, P. 1996. Unconstrained evolution and hard consequences. In E. Sanchez and M. Tomassini (eds.), *Towards Evolvable Hardware*, vol. 1062 of *Lecture Notes in Computer Science*, pp 136–165. Springer-Verlag, Heidelberg.

Toffoli, T. 1977. *Cellular automata mechanics*. Technical Report 208, Comp. Comm. Sci. Dept., The University of Michigan.

Toffoli, T. 1980. Reversible computing. In J. W. De Bakker and J. Van Leeuwen (eds.), *Automata, Languages and Programming*, pp 632–644. Springer-Verlag.

Toffoli, T. 1984. Cellular automata as an alternative to (rather than an approximation of) differential equations in modeling physics. *Physica D*, vol. 10, pp 117–127.

Toffoli, T. and Margolus, N. 1987. *Cellular Automata Machines*. The MIT Press, Cambridge, Massachusetts.

Tomassini, M. 1993. The parallel genetic cellular automata: Application to global function optimization. In R. F. Albrecht, C. R. Reeves, and N. C. Steele (eds.), *Proceedings of the International Conference on Artificial Neural Networks and Genetic Algorithms*, pp 385–391. Springer-Verlag.

Tomassini, M. 1995. A survey of genetic algorithms. In D. Stauffer (ed.), *Annual Reviews of Computational Physics*, vol. III, pp 87–118. World Scientific, Singapore. Also available as: Technical Report 95/137, Department of Computer Science, Swiss Federal Institute of Technology, Lausanne, Switzerland, July, 1995.

Tomassini, M. 1996. Evolutionary algorithms. In E. Sanchez and M. Tomassini (eds.), *Towards Evolvable Hardware*, vol. 1062 of *Lecture Notes in Computer Science*, pp 19–47. Springer-Verlag, Heidelberg.

Vichniac, G. 1984. Simulating physics with cellular automata. *Physica D*, vol. 10, pp 96–115.

Vichniac, G. Y., Tamayo, P., and Hartman, H. 1986. Annealed and quenched inhomogeneous cellular automata. *Journal of Statistical Physics*, vol. 45, pp 875–883.

von Neumann, J. 1966. *Theory of Self-Reproducing Automata*. University of Illinois Press, Illinois. Edited and completed by A. W. Burks.

Wolfram, S. 1983. Statistical mechanics of cellular automata. *Reviews of Modern Physics*, vol. 55, no. 3, pp 601–644.

Wolfram, S. 1984a. Cellular automata as models of complexity. *Nature*, vol. 311, pp 419–424.

Wolfram, S. 1984b. Universality and complexity in cellular automata. *Physica D*, vol. 10, pp 1–35.

Wolfram, S. 1986. Random sequence generation by cellular automata. *Advances in Applied Mathematics*, vol. 7, pp 123–169.

Yager, R. R. and Zadeh, L. A. 1994. *Fuzzy Sets, Neural Networks, and Soft Computing*. Van Nostrand Reinhold, New York.

Index

God appears and God is light
To those poor souls who dwell in night,
But does a human form display
To those who dwell in realms of day.

William Blake, *Auguries of Innocence*

Lecture Notes in Computer Science

For information about Vols. 1–1133

please contact your bookseller or Springer-Verlag